"I will always believe she deliberately tried to span the chasm between our different species—between our different worlds. She failed, yet it was not total failure. So long as I live I shall hear the echoes of that haunting cry . . . I will hear those echoes even if the day should come when none of her nation is left alive in the desecrated seas, and the voices of the great whales have been silenced forever."

"A POIGNANT AND TRAGIC PARABLE FOR OUR TIMES . . . A POWERFUL AND ANGUISHED BOOK."

—*Toronto Globe and Mail*

A WHALE FOR THE KILLING
by
FARLEY MOWAT
Author of
The Boat Who Wouldn't Float,
The Dog Who Wouldn't Be
and *Never Cry Wolf*

ALSO BY

FARLEY MOWAT

A WHALE FOR THE KILLING

Farley Mowat

SEAL BOOKS
McClelland-Bantam, Inc.
Toronto

This edition contains the complete text
of the original hardcover edition.
NOT ONE WORD HAS BEEN OMITTED.

A WHALE FOR THE KILLING
A Seal Book / published by arrangement with
McClelland and Stewart Limited

PUBLISHING HISTORY
McClelland and Stewart edition published June 1972
Seal edition / April 1978

ACKNOWLEDGEMENTS
We wish to thank Holt, Rinehart and Winston, Inc., New York, for
permission to reprint the following:
Passage from The Outermost House by Henry Beston.
Copyright 1928, 1949, © 1956 by Henry Beston.
"The Secret Sits" from The Poetry of Robert Frost edited by Edward
Connery Lathem. Copyright 1942 by Robert Frost.
Copyright © 1969 by Holt, Rinehart and Winston, Inc.
Copyright © 1970 by Lesley Frost Ballantine.

Front cover photograph by E. R. Degginer
courtesy of Bruce Coleman, Inc.
Back cover photograph by John de Visser.

Maps by James Loates

ISBN 0-7704-2331-0

PRINTED IN CANADA

COVER PRINTED IN U.S.A.

U   17   16   15   14   13   12

I wish to thank Peter Davison and Angus Mowat who have helped me with this book more than I can say.

"Remote from universal nature, and living by complicated artifice, man in civilization surveys the creature through the glass of his knowledge and sees thereby a feather magnified and the whole image in distortion. We patronize them for their incompleteness, for their tragic fate of having taken form so far below ourselves. And therein we err, and greatly err. For the animal shall not be measured by man. In a world older and more complete than ours they move finished and complete, gifted with extensions of the senses we have lost or never attained, living by voices we shall never hear. They are not brethren, they are not underlings; they are other nations, caught with ourselves in the net of life and time, fellow prisoners of the splendour and travail of the earth."

HENRY BESTON

THE SECRET SITS

We dance round in a ring and suppose,
But the Secret sits in the middle and knows.

ROBERT FROST

The Outport of Burgeo
and the Surrounding Region

The Island of Newfoundland

Little Barasway

The Ha Ha
Richards Head
Aldridges Head
Greenhill Island
ALDRIDGES POND
Fish Island
Fish Plant
Short Reach
BURGEO
(see opposite page)
Grandy Island
The Harbour
Smalls Island
Messers Cove
Firby Cove
Muddy Hole
Boar Island
The Sandbanks
Messers Head
Franks Island
Eclipse Island
Hunt's Island
Hug My Dug Island
The Burgéo Archipelago
Recontre Island

Scale of miles
0   1   2

Williamsport
White Bay
Notre Dame Bay
Botwood
Gander
Corner Brooks
BURGEO
Grand Falls
Stephenville
Bonavista Bay
Trinity Bay
Dildo
Marystown
Port aux Basques
St. John
St. Albans
St. Pierre et Miquelon
Grand Bank
Burin
Placentia Bay
Argentia Bay
St. Marys Bay
Ferryland
(see accompanying map)

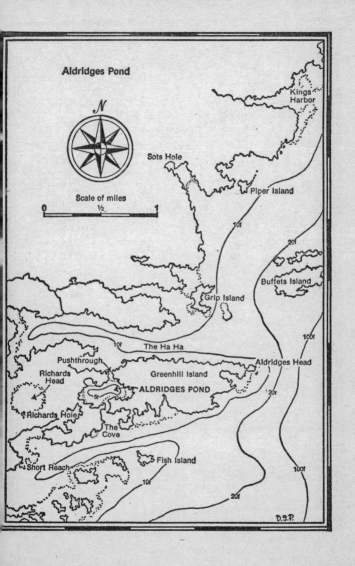

Chapter 1

A torment of sooty cloud scudded out of the mountainous barrens of southeastern Newfoundland. Harried by a furious nor'easter, eddies of sand-sharp snow beat against the town of Port Aux Basques; an unlovely cluster of wooden buildings sprawled across a bed of cold rock and colder muskeg. White frost-smoke swirled up from the waters of the harbour to marry the cloud wrack and go streaming out across Cabot Strait toward the looming cliffs of Cape Breton and the mainland of North America.

January deals harshly with Newfoundland. It had just dealt harshly with me and my wife, Claire, and the hundred or so other passengers who had endured the crossing of the Cabot Strait to Port Aux Basques aboard the slab-sided, floating barn of a car ferry, *William Carson*. The passage from North Sydney, in Nova Scotia, normally takes six hours. This time the storm had extended it to twelve, and the *Carson,* savaged by that surging sweep of wind and water, had meanly revenged herself on passengers and cargo. A ten-ton bulldozer, lashed to the deck with half-inch cables, had been pitched right through the steel bulwarks into the green depths. Grey-faced and desolate, most of the passengers lay helplessly asprawl in cabins reeking with the stench of vomit.

When the *Carson* eventually wallowed into Port Aux Basques harbour and managed to get her lines

1

ashore, there was a grateful if unsteady exodus down
her gangplank. Most of the debarking passengers clambered aboard the antiquated coaches of a narrow-
gauge railway which dawdled its way for six hundred
miles to St. John's, the island capital, on the eastern
coast. However, for a score of men, women and children (Claire and I among them) Port Aux Basques
was not the end of the ordeal by sea. Our destinations
were a scattering of sea-girt fishing villages—outports
they are called—thinly spread along the hundreds of
miles of bold, bald headlands and canyoned fiords of
the island's Sou'west Coast. There was only one way to
reach any of these places—the weekly coastal steamer.

She was waiting for us: small, dowdy, dirty; in
sharp contrast to the sham grandeur of the *Carson*.
But, unprepossessing as she looked, the s.s. *Burgeo*
was wise in the ways of the unforgiving world of water.
She was a proper seaboat, not a floating motel. Day in,
day out for more than twenty years, she had shuttled
east and west along that iron shore, furnishing the
physical link between the outports. She also provided
the principal contact with the outside world for some
forty fishing villages which clung between wind and
water to one of the least hospitable coasts on earth.

By 1967 more than half the outports originally
served by the *Burgeo* had been abandoned—"closed
out," as their forcibly uprooted inhabitants described
it. These age-old settlements had become victims of the
cult of Progress even as the *Burgeo* herself was soon
enough to become such a victim. In 1969 she was condemned, although still as sound as ever, and sold for
scrap—an unwanted anachronism from an age now past
and rejected. Left lying at a wharf in St. John's, she was
stripped by souvenir hunters, and the coldness of a dead
ship spread through her. But she was not quite dead.
One dark winter's night, just before the cutting torches
could start eating into her good Scots iron, she committed herself to her own element. So quietly that not
even the watchman knew what she was about, she settled to the bottom of the harbour, there to become a
monumental embarrassment to the authorities and a

remembered heroine to the many thousands of outport people who had known and loved her during the long years of her service.

However, in mid-January of 1967 the *Burgeo* was still very much alive. Her Master, Captain Ro Penney, welcomed us aboard as we scrambled up the gangplank through a burst of driven sleet.

A small, neat, precise man, Skipper Ro was shy of women. He flushed and ducked his head as Claire came aboard.

"Well, me dear, you're back again," he muttered, apparently addressing his own feet. "Nip in out of the wet now. 'Tis dirty weather . . . dirty weather . . ."

He turned more familiarly to me.

"Come on the bridge, Skipper Mowat. We'd best get under way afore this nor'easter busts its guts!"

During the years Claire and I had known the Sou'west Coast we had made at least a dozen voyages with Skipper Ro. We met him first in the gloomy fiords of Bay Despair in 1961 on a day when I brought my own leaky little schooner alongside the *Burgeo* looking for help in repairing my ancient engine. Not only did I get the assistance of the Chief Engineer, Captain Ro himself came aboard my little vessel, having first asked formal permission to do so. He paid me a high compliment by addressing me as "skipper," and he never failed to use the title whenever we met again.

I wish I could still do him equal honour for, like the *Burgeo*, Captain Ro is also gone. During a heavy gale—it was more nearly a hurricane—in Cabot Strait in the spring of 1970, he took the 10,000-ton train ferry *Patrick Morris* out of North Sydney in response to a distress call from a herring seiner. The seiner foundered before the *Morris* could reach her, and while Skipper Ro was trying to recover the body of one of the drowned fishermen, a forty-foot sea stove in the ferry's stern loading door swamped her, and the big ship began to go down. Captain Ro ordered the crew to the boats but three of the engine room crowd could not be found, and Ro Penney refused to leave without them. He was a quiet man, and steadfast to the end.

Skipper Ro tugged at the whistle lanyard and the *Bur-geo*'s throaty voice rang deep and melancholy over the spume-whipped harbour. The lines came in and we backed out into the stream. Once clear of the fairway buoy, the little ship bent to the gale and headed east, holding close in against the looming, snow-hazed land to find what lee there was.

I went below to the old-fashioned dining saloon with its Victorian, leaded glass windows, worn linen tablecloths, and battered but gleaming silverware. Most of the passengers were gathered there, having a mug-up of tea and bread and butter, and yarning companionably, for on the Sou'west Coast everyone knows everyone else, or is at least known to everyone else. Claire was sitting between the owner of a small dragger and his dumpy, jovial wife. I joined them. The nor'easter screamed in the top-hamper and the old reciprocating steam engine thumped its steady, heavy heartbeat underfoot as we listened to the gossip of the coast.

Had we heard that the government was going to close out the settlement of Grey River? The dragger owner snorted into his cup of tea. "Hah! By the Lard Jasus, they fellers in St. John's is goin' to find they needs a full cargo o' dynamite to shift Grey River. Aye, and I don't say as even that'll shift they people!"

Fish landings had been down. "Entirely too starmy all the fall months. Shore fishermen can't hardly git out at all. Even us fellers onto the draggers, we has to spend the best part of our sea time battened down or runnin' for shelter."

But there were compensations. "Niver did see such a toime for caribou. I tell ye, me son, they's thicker'n flies on a fish flake, and coming right down to the landwash to pick away at the kelp. Oh, yiss bye, they's lots o' country meat on the go!"

He smacked his lips and winked at his wife who promptly took up the tale.

"They got the new school open to Ramea, me dears. Yiss, an Lucy Fenelly, belongs to Mosquito Harbour, got a new baby, and her man away working on the mainland these past ten months! And that young student preacher, come just afore you folks went off, he only stayed long enough to christen the child and then he fair flew off the coast. I don't say 'twas his fault entirely. Lucy's got thirteen youngsters now, and they's none of 'em looks no more like her man than I does meself."

Over the third cup of tea, the dragger skipper, as an act of politeness, asked where we had been.

"Europe," I told him, and added with something of the self-conscious pride of a world traveller, "and Russia. Moscow first, and then right through Siberia as far as the Pacific and the Arctic Coast."

"Roosia, eh? Yiss . . . well now, you'll be some glad to be getting home to Burgeo . . . Me dear man, they's some glut of herring on the coast this winter. Nothin's been seen the like of it for fifty year. . . ."

Burgeo, our destination and the place from which the little steamer took her name, was the largest settlement on the coast, and, through the preceding five years, it had been our home. Now, after six months of kaleidoscopic experiences and exhausting travel, we were yearning for this homecoming with its promise of surcease from the grinding irritations of technological civilizations.

Burgeo lies ninety miles to the east of Port Aux Basques on a shore of such formidable aspect that it remains little known except to the scattering of fishermen and seamen who are its human inhabitants. The Sou'west Coast faces a vast sweep of waters rolling all the way up from the South Atlantic. It is a rare day when this oceanic plain lies quiet. Throughout most of the year onshore gales drive their thundering seas against granite cliffs which rise inland to a high, barren plateau, the home of the caribou and arctic hares, the ptarmigan . . . and not much else.

Lying offshore from the fiord-riven cliffs are clusters of low islands, many of them sea-swept; and seeded among these, like dragon's teeth, lie innumerable underwater reefs and rocks which the coast dwellers call—with chilling simplicity—"sunkers." The number of ships they have wrecked is legion and even in these days of electronic navigational magic, they remain a thing of terror on black and stormswept nights, or when the corpse blanket of fog smothers land and sea alike.

The Burgeo Islands comprise one such cluster. They were "discovered" by Western man in 1520 by a Portuguse explorer, Joaz Alvarez Fagundez. He called the archipelago *"Ilhas Dos Onze Mill Vierges"* in tribute to St. Ursula of Cologne who, in the 14th century, with a naiveté which must be unique in human annals, led 11,000 virgins against the heathens in the Holy Land. Fagundez may have had a sardonic sense of humour for if those wind-swept, rocky islands, surrounded by foaming reefs, were not precisely virginal, they most assuredly were barren.

But the seas around the islands were anything *but* barren. They teemed with life. Seals, whales, even walrus, lived in multitudes in the plankton-rich waters along the abrupt coasts and over the off-lying banks. As for fish! Salmon, cod, halibut, haddock, sole and a dozen other species were so abundant that men standing on shore could spear them by the boatload. Although a hellish place in bad weather, the Sou'west Coast had good harbours, and brave men who would take a harvest from the sea could do so here if they dared the risks.

From Fagundez' time, and doubtless long before, Europeans had dared. By the beginning of the 16th century Basque whalers were on the coast, harpooning leviathan off shore and hauling the giant corpses to the land where the blubber could be rendered into oil. Traces of their tryworks still remain. The French were not long after them. They built summer codfishing stations and, over the years, runaways (deserters from the fishing vessels) hid in the more remote coves and sea-gulches. Here they lived a life almost as primitive as that of the Mic Mac Indians from Nova Scotia who replaced Newfoundland's aboriginal Beothuks when these were hunted to extinction by the encroaching Europeans. The French bred with the Mic Macs, and when a trickle of English and Irish runaways, fleeing the slave ship conditions of the English fishing fleets which came annually to eastern and northeastern Newfoundland, also began to drift along the coast, they were in turn absorbed by the earlier arrivals, and a new breed of men was born.

They were hard and self-sufficient people, as they had need to be in order to survive at all. Because they were outlaws, they dared form no large communities. Dispersal in small groups was also necessary because they possessed only oared boats, in which they could not venture far from home; and too many fishermen working in one place meant crowded grounds.

They clung, limpet-like, to this rock-walled rim of ocean, one or two families together, wherever they could find a toe-hold for their cabins and shelter for their boats. By the late 19th century there were over eighty such clusters of humanity along the Sou'west Coast. Each consisted of from half a dozen to a score of square, two-storey frame houses hugging the foreshore of some stony little hole-in-the-wall where coveys of lean dories and fat-bellied trap skiffs floated at their collars like resting seabirds.

Clinging to the landwash, often at the very foot of a towering cliff, these sparse encrustations of human life were separated one from the other by many miles of unquiet waters, yet united by the sea which was the peoples' livelihood . . . by the sea which was their highway ... by the sea which was their mistress and their master ... the giver, and the takeraway.

Inland, the treeless granite hills rolled starkly naked, but in some of the river gorges there were stands of spruce and larch, and the outport people fled to these protected places during the white months of winter, living in log "tilts" until spring sent them back to the calling sea once more.

It was a rock-hard land, and an ice-cold sea, and together they winnowed the human seed through generations of adversity until the survivors themselves partook of the primal strength of rock and ocean.

Life is easier now, but they are still a breed apart. As late as 1950 they knew little and cared less about the new breed of technological men who had come to dominate the planet. They continued to live in their own time and their own way; and their rhythm was the rhythm of the natural world.

When, in 1957, I first visited the Sou'west Coast, men and boys were still fishing in open, seventeen-foot dories in winter weather of such severity that their mittens often froze to the oars. Some had larger boats driven by antique single-cylinder engines, but these were still open to the sea and sky. Almost all the fishermen brought their cod home to their own spruce-pole wharves, called "stages," and split them in their own fish sheds or "stores." Women and girls still spread the split and salted fish to dry on spidery wooden scaffolds known as "flakes." Salt cod was still the main product of the coast, as it had been for better than three hundred years.

They were truly people out of time, but it was not that alone which drew me to them. Being a people to whom adversity was natural, they had retained a remarkable capacity for tolerance of other human beings, together with qualities of generosity toward one another and toward strangers in their midst which surpassed anything I had ever known before except, perhaps, among the Eskimos. They were the best of people, and I promised myself that one day I would come and live among them and escape from the increasingly mechanistic mainland world with its March Hare preoccupation with witless production for mindless consumption; its disruptive infatuation with change for its own sake; its idiot dedication to the bitch goddess, Progress.

In 1961 I did return, blundering along that fearsome coast in my decrepit little schooner; staying afloat and alive not due to the grace of God, but due to the grace of the outport men who, with a subtlety which enabled me to save face, saw to it that I did not pay the usual price demanded by a harsh environment of fools and amateurs.

The following summer Claire and I coasted westward out of Bay Despair and, by summer's end, neither of us felt any real wish to return to mainland Canada. We began to consider the possibility of roosting for the winter in some small community along the Sou'west Coast, but by late August we had made no decision and were still aimlessly drifting westward.

We were approaching Burgeo, but had no intention of putting in there, for we had heard that the construction of a modern fish-packing plant had radically changed the outport nature of that place. Instead our course was set for the tiny village of Grand Bruit, some miles farther west. However, as we came abeam of Boar Island, which marks the entrance to the intricate maze of runs and tickles between the Burgeo Islands, our engine failed. The runs seemed too hazardous to attempt under sail alone so we reluctantly put in to Burgeo for repairs. By the time the schooner was again fit for sea, the weather had turned foul and we were harbour-bound.

Burgeo, or at any rate the eastern part of it where we lay moored, was not a thing of beauty. The fish plant dominated the scene. The roar of its diesel generators deafened us, and the stench of the place was an abomination under God. Nevertheless, most of the people we met seemed relatively uncontaminated by the arrival in their midst of the industrial age, and they were friendly and helpful. One morning we casually mentioned to one of them that we were thinking of wintering somewhere on the Atlantic seaboard, and almost instantly found ourselves being whisked off to the western end of the attenuated settlement to a little, semi-isolated community called Messers Cove. Even at first glance Messers seemed like everything an outport should be, and nothing it should not be. It consisted of fourteen families of inshore fishermen whose gaily painted houses lined the shores of a snug little harbour. Perched high on a great granite boulder was a small, half-completed, white wooden bungalow whose windows looked south over the islands and beyond to the endless sweep of open ocean.

The little house was for sale.

We offered to rent it but the owner, a young

fisherman who wished to become an hourly labourer
at the fish plant, refused. He would only sell. The season was late, and there was nowhere else in the world
we particularly wanted to be. We liked Messers Cove
and its people. We bought the house.

Captain Penney laid his ship sweetly alongside Burgeo's snow-dusted government wharf. He waved good-
bye from the wing of the bridge as we came down the
gangplank toward a knot of people who had gathered,
as generations before them had habitually gathered,
to meet the coastal boat. An agile little man in his late
thirites separated himself from the group and trotted up
to us. His dark, sharp-featured face was lit by a smile
of welcome.

"Got your wire you was on the boat. Was a
wonnerful glitter starm last week and the road's not
fitten for nothin' only young 'uns on skates. So I
brung me dory for to fetch you home. Where's your
gear to? I'll get it for you."

Simeon Spencer, proprietor of a tiny general store
which occupied the back kitchen of his house, was our
closest neighbour at Messers, and perhaps our closest
friend. With great solicitude he saw us aboard his
powered dory, stowed our luggage, and cast off.

It was a bitterly cold evening and, despite the
wind-lop, a rubbery film of cat-ice was forming over
the runs between the islands leading past Firby Harbour, Ship Dock and Muddy Hole, and westward to
our own cove—Messers. Spray froze on the backs of
our coats as we hunched forward on the thwart facing
the diminutive figure of Sim who stood erect and bird-
like in the stern. Unexpectedly, he shoved the rudder
hard down and the dory lay over so sharply that Claire
and I slid sideways on the thwart. Sim was waving an
arm seaward, and over the clatter of the engine I could
just make out the word he shouted.

"Whales!"

He was pointing toward Longboat Rocks, a line of
black reefs glistening with the salt seas breaking over
them. Just beyond the Longboats was something else,

also black and glistening, that surged slickly into view, then sank smoothly from sight leaving behind it a plume of fine mist which was quickly blown away by the nor'easter.

That blurred, elusive glimpse of one of the Great Ones of the oceans was a fine homecoming gift. I have always been fascinated by the mysterious lives of the non-human animals who share this world with us, but until I went to live on the Sou'west Coast the mystery of the whale had scarcely come my way, although it is perhaps the greatest mystery of all. Burgeo had given me the chance to approach that mystery when, each winter, a little group of Fin Whales took up seasonal residence for a few months in the waters of the archipelago.

Sim had turned the dory into Messers Cove. He cut the engine as we approached his stage.

"When did the whales come back?" I asked as the dory drifted toward the wharf.

"Like always. First part o' December . . . along of the herring scull. They's five . . . maybe six cruising midst the islands, and they's the biggest kind! . . . Here now, Missis, watch the ice!"

Together we helped Claire up the slippery front of the stage and in the excitement of this long-anticipated arrival the whales were temporarily put out of mind. None of us could then have guessed the momentous changes in all our lives which was to follow from their presence.

All the lights in our house were burning. When we stamped across the storm porch and entered the kitchen we found the stoves all roaring. Sim's fourteen-year-old daughter, Dorothy, had swept and polished every inch of every room. Old Mrs. Harvey had sent over two loaves of homemade bread, still oven-hot. And bubbling fragrantly in a pot on the oil range was a boiled dinner: cabbage, turnips, potatoes, onions, salt beef and moose meat.

The little house which had stood empty for six months was as warm and welcoming as if we had never left it. Our delight was also Sim's delight as he sat unobtrusively nursing a glass of rum in front of the Franklin stove. His self-imposed duty was not yet completed. It was now his task to bring us up-to-date on the really vital events which had taken place during our absence.

"They was talk of a strike down to the plant, but the owner, he put an end to it right quick. Said he'd close her up and move clear of Burgeo and leff the people lump it if they didn't care to work to his wish. Don't know where he figgered on going, but they's plenty folk here could name a place proper for the likes of he! . . . Curt Bungay, he's bought a new boat, a longliner from down Parsons Harbour way . . . Your schooner's hauled up clear, and Joe, he found her leak . . . one of 'em anyhow . . . They's a new nurse to the hospital; chinee they *says* she is, but she's good for it. Comes out the dirtiest sort of weather to take care of folks . . . Good run o' fish this winter, but nothin fit to call a price . . ."

And so it went, in rapid fire bursts, until a decent time having been allowed for us to have our dinner and settle in a bit, other visitors began clumping over the porch and unceremoniously letting themselves into the kitchen. A bevy of teen-age girls led the way. They sat in a tidy row on the kitchen day-bed and said nothing; only beamed, giggled and nodded as we tried to make conversation with them. Then, with familiar uproar, Albert, our big, black water dog, came gallumphing into the kitchen bearing a dried codfish as a homecoming present. Close behind him came his, and our, friend, eighty-year-old Uncle Job; gnome-like, ebullient and grinning thirstily as he spotted the rum bottle on the kitchen table. During our absence Albert had lived with Uncle Job and his wife and, according to what Sim told us later, dog and man had spent most of their time engaged in acrimonious debate as to whether they would, or would not, go fishing, go for a walk, go for a swim, go to bed, or get up in the morning. They were both inveterate argufiers.

"He'll miss that dog some bad," Sim said. "Wuss'n

he'd miss his woman. *She* never says a word, and it drives he fair wild when he can't get an argument."

Rather reluctantly Albert delivered over the cod —a fine one it was too, for Uncle Job was one of the few Burgeo people who still knew and practised the age-old art of drying and salting fish. Albert then sniffed in a perfunctory way at our luggage with its stickers from Irkutsk, Omsk, Tbilisi and other exotic places, barged into the living room, climbed up on the sofa, grunted once or twice, and went to sleep.

The homecoming was complete.

Chapter 2

The storm blew itself out that night and in the morning the sun shone white and harsh upon the ice-capped offer islands. The sea had an ebony sheen to match the glitter left on the land by the recent sleet storm. Taking advantage of what would surely be an all-too-brief spell of fine weather, the inshore fishermen—those who fished the local grounds with gill nets, trawls and handlines—were early at work, their small boats no more than distant motes on the metallic mirror of the sea.

Leaving Claire to get on with the unpacking as best she might under the solemn eyes of the neighbourhood children who drifted in and out of the kitchen like voiceless wraiths, I took Albert and set out to collect the mail. Since the trail (it could only be called a road out of courtesy) leading to the eastern end of Burgeo was still impassable because of its skin of ice, I borrowed Sim's dory and took the water route instead.

It was marvellously exhilarating to be on the sea. The day was calm and the sky blindingly clear. As usual, Albert rode standing up in the bows, leaning far forward so that he looked like a pagan totem. When occasional dovekies and sea pigeons rose heavily ahead of us and skittered off to the sides, he eyed them with disdain.

We puttered along close under the land—close

enough to sniff the iodine tang of kelp exposed at ebb tide; close enough to recognize and be recognized by people in the shoreside houses. Sim Spencer's wife, busy hanging out the wash, gave us a welcoming wave. Josh Harvey's calico dog, Jumbo, barked insults at Albert from a stage head, and Albert richly returned them. I shouted a greeting at Uncle Matt Fudge, ninety-one years old, but still young enough to be mending a cod net in the sunny shelter of his grandson's fish store. (Everyone above the age of fifty in an outport is known to everybody else as "Uncle," or as "Aunt.")

From Messers as far as Frank Island the shore we were coasting was encrusted by a fringe of sturdy houses which had been occupied for generations. They were the "old" Burgeo, and they were a joy to behold. They looked married to the rock on which they stood, a neat but unobtrusive frieze along the shore. Yet each home stood in quiet singularity. No two faced in the same direction. No two were on the same level nor were of quite the same construction, even though most were of the venerable two-storey, low-peaked pattern common to Newfoundland outports. They gave the impression of a kind of unpretentious and nonaggressive independence. However, when the dory slipped through the tickle north of Frank Island, the scene changed abruptly. This was where the New Burgeo, creation of modern times, began.

In 1949, when Newfoundland became a reluctant part of Canada, Burgeo was not a single community but a patchwork of dispersed little settlements. Along the shoreline of Grandy Island—which dominates the archipelago as an anchored battleship dominates a fleet of lesser craft—lay the hamlets of Messers Cove, Muddy Hole, Firby Cove, Samways, The Harbour, and The Reach. Clinging to the offlying and much smaller islands, or strung along the adjacent mainland shores, lay Seal Brook, Kings Harbour, Our Harbour, Hunts Island, Sandbanks and Upper Burgeo. Although none of these held more than two dozen families, each had preserved its own identity through a century and more.

Confederation with Canada put an end to all that. In 1948 Newfoundland was still nominally a self-governing Dominion in the British Empire; but in 1949, goaded and harried by a messianic little man named Joseph Smallwood (some called him satanic), Newfoundland was stampeded into joining Canada. Smallwood won the decisive vote by the slimmest of margins as islanders of all classes fought desperately for the retention of their independence, impoverished as it was. For these dissenters, independence was of greater worth than flash prosperity. Smallwood, on the other hand, regarded independence as an insufferable barrier to progress. Most Newfoundlanders, he once contemptuously said, did not know what was good for them and would have to be hauled, kicking and screaming, into the 20th century. He was just the man to do the hauling.

He became the island's first provincial Premier and during the next twenty-two years ran Newfoundland almost single-handedly, according to his personal concept of what was good for it. It was a simple concept: industrialize at all costs. This meant that all the island's mineral, forest and human resources were to be made available, virtually as gifts, to any foreign industrial entrepreneurs who would agree to exploit them. Smallwood demanded that Newfoundland turn its back on the ocean which had nurtured the islanders through so many centuries.

"Haul up your boats . . . burn your fishing gear!" he shouted during one impassioned speech directed at the outport men. "There'll be three jobs ashore for every one of you. You'll never have to go fishing again!"

Many believed him, for he was a persuasive demagogue, and he had the silver tongue.

One of the first hurdles he had to overcome was to find means of concentrating the "labour resources" (by which term he described the people of the outports) who were dispersed in about thirteen hundred little communities scattered along some five thousand miles of coastline. Smallwood's solution was "Centralization" which, translated, meant the forced and calculated merger of the outports so that labour pools could be formed from the transported occupants. The methods used to destroy the small outports were devious, usually deceitful, sometimes brutal . . . and almost always effective.

Along the Sou'west Coast the "moving fever," implanted and cultured by the Smallwood men, soon began to take effect. One by one the outports sickened and died. Even in the Burgeo archipelago, where everyone already lived within a four-mile radius of everyone else, the fever raged with such fury that in a few years all the off-lying communities had moved to Grandy Island.

Although Smallwood rejected the sea and scorned fishing as a way of life, his was not an absolute rejection. Even in his most sanguine dreams he appears to have realized that there were parts of Newfoundland which could not be turned into facsimiles of Detroit or Hamilton. The Sou'west Coast was such an area. Smallwood's answer to how best to exploit its labour potential was to heavily subsidize construction of a fish-freezing plant on Short Reach, at the east end of Grandy Island. This plant was "sold" for a ridiculously small sum to the son of a St. John's merchant prince who found himself in the happy position of being able to pay what he chose for labour while setting his own price for the fish he bought.

There were some initial difficulties. Not many men could be persuaded to abandon their way of life in order to become wage employees at as little as ten or twenty dollars a week. However, as the people of the neighbouring outports began to converge on the "growth centre" of Burgeo, a surplus labour force developed. It consisted of people who had always hated the very thought of welfare—the dole, they called it—and were willing to work at almost anything, for almost any wage, rather than accept relief.

The heaviest concentration of immigrants settled close to the fish plant. When all the available shoreline was occupied, newcomers were forced to build away from the sea on barren rocky ridges or on peaty muskegs. They built hurriedly and, contrary to their wont, many built badly. They had no money with which to buy materials and, since they were wage slaves, they had no time to do as their fathers had done and go into the country to cut and whip-saw their own lumber. All too many of the new residents, who had been forced or deluded into abandoning comfortable and well-built houses in the now deserted outports, were reduced to living in unsightly shacks. These proliferated until they produced the first true fruits of Centralization . . . the Sou'west Coast's first slum.

The eastern end of Grandy Island turned into a wasteland of rusting cans, broken bottles, spilled garbage, and human sewage. The surrounding waters were further defiled by the vast volume of effluvia from the fish plant which discharged all its wastes and offal directly into Short Reach. Much of the shoreline was befouled by a belt of black sticky muck several inches thick and six to ten feet broad which, particularly at low tide, stank to high heaven.

Apart from the physical degradation of what had once been a wholesome and natural environment, centralization also degraded the people who were its victims. The delicate interdependence of give and take was disrupted. As Sim Spencer once explained it to me:

"Afore the plant come here, every little place was on its own. Every settlement was like a family; and all the families, all along the coast, they got along good, never rubbed nobody the wrong way. Every man looked after his own, but he looked to his neighbour when times was hard, and for certain sure he'd be quick enough to give a hand when anyone else was needy.

"Now that's all gone abroad. Shovin' all hands into one big lobster pot done something to the people. Started fillin' them up with badness. Turned them one agin the other. Started them wheening and growling like a pack of crackie dogs penned up on some little pick of an island.

"These times, everyone's jealous of t'other, and 'twas never that way before. People is uncontented. They don't want nobody to get nothin' unless they gits it too, *and* more of it. Truth to tell, people is turning right hateful in Burgeo, these times."

The independent people, and the egalitarian way of life, could not survive the tides of change as more and more people came to Burgeo from more and more "closed-out" outports. There was not nearly enough wage work to go around. The new men could not fish successfully for themselves because the local grounds were foreign to them and were by now seriously over-crowded and overfished as well. Consequently, scores of men, both young and old, were forced to leave their families behind and seek work, not only outside Burgeo, but outside Newfoundland where the grandiose industrial schemes of Premier Smallwood had come to nothing. Some men worked seasonally in Nova Scotia; others spent eight months of each year manning Great Lakes freighters—eight months away from home.

As if this were not enough, compulsive consumer-ism, the universal sickness of modern society, infected the dispossessed outport people. Men, women and children who had never cared much for material posses-sions became greedily acquisitive. They began to thumb avidly through the shiny mail order catalogues. The solid, hand-made furniture they had brought with them from the outports was now "condemned"—thrown over the ends of the stages to float away on the tides. It was replaced with chrome and arborite.

The plant owner, ever anxious to develop con-sumer attitudes and appetites, opened a supermarket. Some of the befuddled victims of the new disease ac-tually bought television sets even though there was no transmitter which could reach them.

Electricity and roads spread their web through the increasingly congested maze of houses. Two miles of incredibly rough trails were hewn from the living rock and, in 1962, the first two cars were unloaded from the coastal boat. They met a happy fate a few days later when they collided head-on and both were reduced to junk.

During the five years Claire and I lived in Bur-geo, Progress made further stunning strides. By 1967 there were thirty-nine cars and trucks rattling themselves to pieces on the stony goat tracks which led nowhere, and never would. The first snowmobile had gone snarling out into the barrens, where it fell into a crevice; but the following year saw five more on order. The non-returnable pop and beer bottle arrived. On sunny summer days the rocks which made up most of the physical surroundings gleamed and glittered in fairy colours from the layers of glass shards which littered them. In 1961 there had been no welfare officer and no unemployment. But by 1967 Burgeo had these modern advantages as well as a new fish meal reduction plant to spread its oily, nauseous fumes like a miasma over the entire community.

It had also acquired a town council. It had a Mayor—the owner of the fish plant. He was a man after Smallwood's own heart, and one whose motto might well have been: "What's good for me is good for Burgeo." Since the town council was chosen for the most part from among his employees, or his sycophants, he encountered little opposition.

There was also a fine new school built to mainland standards and staffed with "modern" teachers who were skilled at denigrating the old ways, rejecting the past, and arousing in their students the lust for the golden dreams of the industrial millennium.

The people of the Sou'west Coast, and of Burgeo in particular, were "hauled into the 20th century" so speedily that few of them had any understanding of what was happening to them. The age-old patterns of their lives collapsed in rapid succession. The inner certainties which had sustained them in past generations were evaporating like water spilled on a red-hot stove. But not all Burgeo people failed to grasp the significance of what was happening to them. Some of them understood.

There was Uncle Bert, for example. He lived with his "Woman," as he always referred to his wife, in a tiny but impeccably neat little house a stone's throw from us at Messers. In the evenings Uncle Bert would sit at his oilcloth-covered kitchen table listening disdainfully to a squalling transistor radio as it yammered out the daily tale of hate and horror, of suffering and disaster, which it offered as the news of the world.

When the tale was told, Uncle Bert would switch off the radio, pour himself half a glass of straight alcohol (smuggled in from the offshore French island of St. Pierre), top it up with boiling water, add a spoonful of sugar, and toss the mixture down in a couple of gulps. Then, bald head shining with sweat from the effect of the "alky," big, twisted hands gesticulating in the light of the oil lamp, he would bellow out his derision.

"By the Lard livin' Jasus, dem mainland fellers is gone altogether foolish! Foolish as a cut cat, me son! And de great joke onto it . . . dey don't *know* it! Dey got to tinker wit' *every* goddamn t'ing dere is . . . and everyt'ing dey tinkers wit' goes wrong! And dat, me darlin' man, *dat's* what dey calls *pro*-gress!

"Dey says dey's makin' a heaven on dis eart' for we. But de troot onto it is, dey's headin' dereselves and all of we for hell, in a hoopin hurry-all. Smart? Oh yiss, dey do believe dey's de smartest t'ings God put on dis old eart' . . . dem politicians and dem scientists, and all dem fine, big-moneyed fellows. But I'm tellin' ye, byes, de codfish and de caribou, dey's ten t'ousan' times smarter in de head. *Dey* got de sense to lave well enough be. *Dey'll* niver blow up de world; no, nor pizzen us all to deat' . . . Bejasus, byes, I t'ink de healt's gone right out of we!"

It was not gone out of Uncle Bert. At seventy-six he still went fishing single-handed in his dory, winter or summer, whenever the fish "were on the go," even though he and his Woman between them had an ample income from their oldage pensions.

"I goes because I got to fish! 'Tis what I wants to do. 'Tis out dere, on de sea, dat's where I wants to be. Dat's where I knows *what* I is, and *who* I is . . . a damn good man, and I ain't feared to tell it!"

He minced no words in describing what he thought about the new way of life in Burgeo. "Dem poor bloody bastards as works to de plant! Workin' for wages and t'inking deyselves some lucky! I'll tell ye what dey is, me sons . . . dey's slaves. No better dan slaves! And de wors kind of slaves, 'cause dey's *grateful* for a chance to work for de owner in dat stinkin' shithole down to t'Reach for de rest of dere lives, so's dey can buy some goddamn t'ing is no more use to dey dan legs is to a fish! Cars, byes, and tely-veesion. Sewry poipes and houtboard ingines! Dem fellers don't know who dey is no more . . . dey only knows what dey wants! And what dey wants is enough to bloat a hog until he busts. Dey's comin' out no different from all de people up along in Canada and America. Dey wants it *all!* And so dey gets de chance, dey's goin' to *take* it all. And den, by de Lard livin' Jasus, dey's goin' to choke dereselves to deat' on dere own vomit!

"Sometimes I tinks de lot of 'em should be in de Mental.* Whatever dat plant owner fellow tells 'em, dey eats down as slick as cold boiled pork. But he don't give a bloody damn for dey. One marnin' dey'll wake up wit' nothin' left 'cept a razor for to cut dere t'roats."†

Preferring not to face the stench of Firby Cove, I landed at the slightly less malodorous Ship Dock and, leaving Albert to look for rats among the accumulation of garbage on the shore, made my way through a crowded and slovenly collection of shacks and shanties to the brand-new post office which had become the prime symbol of Burgeo's new goals. It was box-square and ultra-modern, built of brick and glass and chrome—the only such building on the entire Sou'west Coast. It had been built during my absence abroad, replacing a cosy, crowded room in one of the old wood-

---

*The St. John's Mental Home, an obsolete institution for the deranged, in St. John's.
†In the late summer of 1971 the workers at the fish plant finally struck for union recognition. This was refused them. When they persisted, the plant owner simply closed down the plant and departed, lock, stock and barrel; abandoning the plant, his job as Mayor, and Burgeo itself.

en houses where the mail had been sorted and distributed for thirty years by Uncle Ted Banfield.

I was going to miss Uncle Ted, who had been forcibly retired when the new post office opened. He knew more about the Sou'west Coast than any man alive and he was free with his knowledge. On cold winter days he was free with his hospitality too. The long walk from Messers against a winter gale was something to chill the blood of an Eskimo, but it was Ted's kindly habit to take you into his kitchen and pour you a warming four-finger glass of rum before delivering up your mail. Banfield's old house had been a place in which to linger, to chat with friends, to hear the local news. The new post office, ultra-sterile under the glare of fluorescent lights (NO DOGS ALLOWED . . . WIPE YOUR FEET BEFORE ENTERING), was a place to go into and get out of again with the utmost speed. I hardly recognized the pallid and harassed-looking young man who thrust my mail at me without uttering a single word. He was Uncle Ted Banfield's son.

Back in the dory I set a return course for home which would take us out and around Eclipse Island and through the deep channel known as Steamer Run. I was hoping to get a glimpse of the whales, but on this fine day they must have gone to sea with the other fishermen, for I saw no sign of them.

Chapter 3

Apart from occasional visits to the post office, Claire and I saw little, by preference, of the eastern end of Burgeo. Messers was our home and we had come to feel very *much* at home there. We had been given tolerance and friendship by all the families whose trim and cared-for houses stood, well spaced, around the rocky rim of the clean little cove. They were people who had lived all of their lives, even as their forbears had done before them, in this place. They had not yet been much affected by the changes which were so rapidly transforming the rest of Burgeo.

During the first week after our return we were visited by most of our neighbours; and warm, welcoming visits they were too.

One of the first to come was Onie Stickland, a sad-faced bachelor of middle age, and one of the few dory-men on the coast still fishing single-handed. Onie brought us a bucket of fresh herring. The herring schools were beginning to run heavily inshore, he told us, and he would be glad to bring us some every day. That about exhausted the conversation since Onie was a listener, not a talker. Shy, gentle, almost pathetic in his eagerness not to be an imposition on us, he was content to sit silently for hours in our house covertly glancing now and again at Claire with an expression of distant adoration.

Simeon Ballard was another frequent visitor.

Heavy-set, bluff-browed and a seaman through and through, he was as loquacious as Onie was reticent. Simeon had been a great wanderer, sailing to the Caribbean in three-masted schooners laden with salt cod, returning with cargoes of salt, molasses and rum. In sail, or in steam, he had visited the great ports in South America, the Mediterranean, and the Baltic. Between voyages he had not been idle, for he had sired nineteen children, seventeen of whom were still alive. Although still in his prime at the time of Confederation, he had hardly gone to sea since then.

He was a soft-spoken and extremely courteous man who made a point of calling me "Skipper" because I owned what may well have been the last schooner in Newfoundland still under sail.

"Them days," he meant before 1949, "they was always eight or ten big schooners belonged to Burgeo. Spring and fall we fished the banks, and summer-times we voyaged foreign. 'Twas a hard life, I suppose, but never seemed that way to we. They was a dozen skipper-men in Burgeo and no place on the seas they couldn't take a vessel. Aye, and bring her home again!

"After Confederation that all went abroad. Seems they fellows in Canada had no use for we. The schooners was laid up to rot and the steamers was mostly sold away. Some of we went at the fishing game, but that began failing too, and so the most of us had to come ashore. 'Twas hard for a man my age—forty I was then—with my master's ticket and all, and still good for it, but no work to be had.

"Aye, 'twas hard enough. But they tells we 'tis all for the best. They tells we a man has a better life working in a factory. Maybe 'tis so, but I thanks my God for the life I had at sea."

One evening well after dark, when not too many people would notice his coming, we had a visit from Uncle Samuel. He was a whipcord little man whose walnut wizened face betokened his Indian blood. He was not much of a fisherman and, indeed, he did not like the sea; but he was a famous countryman—infamous if you listened to the Royal Canadian Mounted Police constable who was Burgeo's arm of the law.

Uncle Samuel was a hunter and trapper. His world was the bleak, wind-eroded barrens which stretched north from the coast for a hundred miles. On this visit he brought us a huge, dripping brown-paper parcel which contained not one, but several feeds of "country meat," the euphemism for illegally killed caribou.

Warming himself generously with my rum, Uncle Samuel talked for hours of the things which were his world. He told us it was a great year for lynx; that scores of the big cats had left the shelter of the distant forests to roam the barrens in pursuit of hares. Moose, he said, were fairly common in the river valleys, but he bemoaned the growing numbers of "sports" (he used the word with supreme contempt) from "down to the Reach" who were using their pay cheques to buy magazine-loading rifles with which to kill every moose they saw.

" 'Tisn't the killing as I minds," Samuel complained. " 'Tis the wicked waste. They fellows now, they'll go into the country and if they spies a calf or a cow they'll kill it quick as they would a bull. Aye, and likely nivver carry a pick of meat back out with them.

" 'Tis not like the old times. Them days if a man was lucky enough to have a gun and shot and powder, and killed a deer, he'd carry ivvery last scrap home to the mouths of his woman and his young'uns. . . ."

Uncle Samuel paused to shake his head in disgust.

" 'Tis hard times when growed men'll take on like they does now. 'Tis something new. They's not a man I growed up alongside would kill more'n him and his folks could use. Now, be Jasus, them fellows down to the plant is all for goin' gunnin' ivvery chance they gets, for ivvery thing as crawls, or walks, or flies. Last week some of they killed two gripes* just for the sport onto it. I never heered of ony man shoot one of *they* before. 'Tis spiteful against nature, 'tis what *I* calls it!"

Although most of our visitors were from Messers,

---

*Bald Eagles.

occasionally people from "the other end" (the term by
which our neighbours obliquely referred to The Harbour and The Reach) also came calling. One afternoon
Albert's stentorian barking brought me to the window. Picking their ways nervously across the footbridge
which joins Messers to the rest of Burgeo were five
strapping-big saddle horses. The riders were the hus-
band-and-wife doctor team from the Burgeo Cottage
Hospital; their two children and their amiable handy-
man whom they liked to call their groom. All save the
"groom" were impeccably dressed in English riding
costumes—jodhpurs, riding crops, hunting jackets and
duck-billed caps. They were followed by two of those
enormous black woolly mammoths which are called
"Newfoundland" dogs but which actually originated in
England a century ago.

The doctors' family was one of two comprising the
Burgeo "aristocracy." As recent immigrants from Britain, the doctors perhaps considered themselves to be
on a higher social plane than the second family, which
was that of the fish plant owner. But both were united
in their determination to impose the social standards
of country gentlefolk on the Burgeo background, and
they competed mercilessly for top billing, using the
tools of conspicuous consumption. Thus when the doctors bought a jet-propelled speedboat which would do
thirty knots, the plant owner responded by purchasing
a cabin cruiser of regal splendour. It was an unfair
competition. The doctors, as salaried employees of the
Provincial Health Service, had to make-do on a fixed
income not much in excess of $35,000.00 a year, but
this was a pittance compared to the income the plant
owner and his wife drew from their several enterprises.

The competition also had its ludicrous side; one
which was not lost on the people of Burgeo. When
the doctors went equestrian and imported two riding
horses, the plant family riposted by buying four thor-
oughbreds. The doctors met this challenge by bringing
in two more horses *and* a Shetland pony. The plant
owner's reply was to import *four* more horses . . .

and a Mexican burro. Then, to settle the matter for good, he added a brace of Peruvian llamas! The doctors gave up and turned the competition into different channels.

"No telling where 'twould have stopped," was Sim Spencer's acid comment. "I don't doubt they people'd have brung in gee-raffs next, and elephants, 'til they'd have been no room left for we!"

After a week at home our lives had settled down to the Messers pace and we were beginning to regain the feeling of tranquillity which was one of the great benefits of outport living. There was time for long walks along the shores, beachcombing; or for treks into the country where we occasionally glimpsed herds of caribou; or to the Barasway, a salt water lagoon surrounded by white sand beaches which were much favoured in summer by children's swimming parties and clam digging excursions.

One sunny afternoon Albert and I crossed the precarious suspension bridge—it was barely three feet wide and swung like a skipping rope in heavy weather—that links Grandy Island to the mainland on the west. Together we climbed the steep slopes of The Head, a massive dome of granite towering two hundred feet above the seas breaking against the encrusting ice.

When we reached the crest it was to find we were not the first to seek this magnificent vantage point, with its horizon-wide sweep of islands and open ocean. Sitting as motionless upon a lip of granite as if he were an extension of the rock was a man whose lean, attenuated figure and hawk-nosed profile belonged to the patriarch of Messers, Uncle Arthur Pink.

Uncle Art was peering out toward Rencontre Island through the tube of a big brass telescope which must have been even older than he was. At seventy-eight, Uncle Art was another fisherman who clung resolutely to the old ways. His boat was an elegant, lap-streak trap skiff which he had built himself and fitted with a thunderous 5 h.p. "make-and-break" engine of almost prehistoric vintage. In this little vessel he would go anywhere in any weather. People used to say of him:

"Dey's nothin' to stop that man! He'd sail to hell and pull the devil's nose if they was fish to be got!"

But Uncle Art was much more than a master fisherman. He was also a man possessed of an amazingly keen and curious mind. Whatever he saw, heard, smelled or touched became a part of his awareness . . . something to be remembered and thought about. All his life he had minutely observed everything upon the sea and much that lay beneath its surface.

"Evening, Uncle Art," I said. (Any time after noon is "evening" on the Sou'west Coast.) "Scunnin' for whales?"

He lowered the glass and gave me his slow smile.

"Aye, skipper. And isn't they some smart? See that herring seiner off yonder? Brand-new iron ship, she is . . . two hunnert tons or more, I'd say. She's got every modern kind of gear for killing herring. I been watching her work. And half a mile inside of her they's a pod of whales. I been watchin' they, too, and I'll wager they be fishing twice as smart as all that fine machinery, and twenty men besides."

He chuckled happily, which seemed a little odd since by all the rules of the game he ought to have been on the side of the fishermen and against any animal which competed with them. However, I knew that Uncle Art was a whale enthusiast. At the age of ten he had begun accompanying his father in a four-oared dory to the dangerous offshore fishery at the Penguin Islands. Here, while handlining for cod, he met his first whales.

" 'Twas a winter fishery them times, and hard enough. The Penguins lies twenty miles offshore. They's nothin' more'n a mess of reefs and sunkers, feather-white with breakers in any kind of a breeze, but the foinest kind of place for cod, and herring too. We'd row out there on a Monday and stay till we'd finished up our grub . . . sometimes ten days. Night-time, and in bad weather, we'd pitch on the rocks under a bit of sailcloth.

"They was t'ousands of the big whales on the coast them times. Companies of them would be fishing herring at the Penguins whilst we was fishing cod. Times we'd be the only boat, but the whales made it seem like we was in the middle of a girt big fleet. They whales never hurted we, and we never hurted they. Many's the toime a right girt bull, five times the length of our dory, would spout so close alongside you could have spit tobaccy down his vent. My old Dad claimed they'd do it a-purpose; a kind of a joke, you understand. We never minded none, for we was in our ileskins anyway.

"And I'll tell you a quare thing. So long as they was on the fishing grounds along of we, I never was afeared of anything; no, nor never felt lonely neither. But after times, when the whales was all done to death, I'd be on the Penguin grounds with nothing livin' to be seen and I'd get a feeling in me belly, like the world was empty. Yiss, me son, I missed them whales when they was gone.

" 'Tis strange. Some folks says as whales is only fish. No bye! They's too smart for fish. *I* don't say as what they's not the smartest creatures in God's ocean."

He paused for a long moment, picked up the telescope and gazed through it.

"Aye . . . and maybe out of it as well."

Chapter 4

Most assuredly whales are not fish, although until a century ago most people, including those who knew them best, their hunters, thought they were. Whales and men trace the same ancient lineage through creatures born in the warm waters of the primal oceans who exiled themselves to the precarious environment of the dry land. They continued to share a common ancestry through the long, slow evolution that began with the amphibians and eventually led to the mammals. But whereas our mammalian fathers stayed ashore, about a hundred million years ago the mammalian forbears of the whales chose to return to the mother of all life—the sea. The descendants of the whale forefathers now number about a hundred species which man, the great cataloguer, has divided into the families of the toothed and the baleen whales.

The toothed whales are the more primitive but the most various, for they include all the porpoises and dolphins, the Sperm, Killer, White Whale, and that unicorn of the sea, the Narwhal. Except for the Sperm, which grows to sixty feet, most of the toothed whales are relatively small, some being less than four feet long.

There are only eleven species of baleen whales, but they rank at the top of the whale's evolutionary tree. About eighteen million years ago, when our own ancestors were abandoning the forests to awkwardly start a new way of life as bipeds on the African savannahs, some of the whales began abandoning teeth in favour of fringed, horn-like plates (baleen) that hang from the roof of the mouth to form a sieve with which the owner strains out of the sea water immense quantities of tiny, shrimp-like creatures, or whole schools of little fishes. It seems a paradox that the largest beast in the world should prey on some of the smallest, but the system works surprisingly well. The proof of the pudding is in the eating, for the baleen whales are the most stupendous animals that ever lived. They include fifty-foot Greys and Seis; sixty-foot Rights and Humpbacks; eighty-foot Finners; and the giant of all time, the Blue, which may grow to a hundred and fifteen feet and weigh almost two hundred tons.

Although they bear a superficial resemblance to fishes, whales have little in common with the scaly tribe. When they returned to the sea they brought with them an intelligence of a radically new order— one that had evolved as a direct consequence of the ferocious difficulties which all terrestrial animals must face in order to survive, and which reached its peak in the mammals. This legacy was shared by the ancestral whales and by the nameless creatures who were to become the progenitors of man.

In the case of *our* forbears, intelligence continued to develop along terrestrial lines in order to cope with the original stringent need which called it forth: the need to survive under competitive and environmental conditions of appalling severity. From this fierce struggle man ultimately emerged with the most highly developed brain of any land animal, and he used it to become the most ruthless and destructive form of life ever to exist. Intellectual supremacy allowed him to dominate all other forms of life, but it also enabled him to escape the restraints—the natural checks and balances—which had prevented any previous species from running hog-wild and becoming a scourge unto life itself.

It was a different story with the whales. When their ancestors returned to the sea it was to an environment which, compared to the land, was positively ami-able. Instead of having to scrabble for survival on the dry, restricted, two-dimensional skin of the planet, split as it was into fragments separated by impassable seas, they returned to the wet, three-dimensional and inter-connected world of waters which surrounds and iso-lates the land-islands. Here they were free to go where and when they pleased. They were restored to a womb world where the climate was more stable; where there was no shortage of food; and where there was no need to occupy or defend territory. Because the ancestral whales returned to this water world endowed with the survival skills so hard-won on land, they were as superior to the old, cold-blooded residents of the seas as time-travellers from some point millions of years in the future might be to us.

Pursuing their advantage over other sea-dwellers, the whales underwent a leisurely evolution extending over many millions of years during which they achieved near-perfect adaptation to the sea environment.

On the other hand, the emerging human stock had to battle desperately for survival in a bitterly rig-orous environment, not only against an array of other animals which were often physically and functionally superior, but against the organized and war-like com-petition of their own species. The human stock would surely have been eliminated if it had not used its de-veloping brain to invent ways of redressing the bal-ance. Faced, as he so often was, with an intolerable climate, man learned to build shelters, use fire, and make clothes. Faced with physically superior animals of other species, *and* with deadly competition from his fellows, he made weapons. Faced with the constant spectre of starvation, he made tools with which to cultivate his own food supplies. Bit by bit he stopped relying on natural evolution to keep him alive and in the race, and came to lean more and more heavily on artificial substitutes. He had invented, and had become a slave to, technology.

Whales never needed a technology. Going back to the sea enabled them to survive successfully as natural beings . . . yet they were beings who, like proto-man, were endowed with a great intellectual potential. What did they do with it . . . with their share of our mutual legacy? We simply do not know. Despite our much-vaunted ability to probe the secrets of the universe, we have so far failed to probe the mystery of the mind of the whale.

Such studies as we have made suggest that the more advanced whales have brains comparable to and perhaps even superior to ours, both in complexity and capacity. It is clear that their power to think has steadily increased, even as ours has, over the millenniums. There can only be one reasonable assumption from all this: whales must use their minds, and use them fully in some direction, in some manner, for some purpose which evades our comprehension. For it is an immutable law of nature that any organ, capability or function, which is not kept well honed by constant use will atrophy and disappear . . . and the brain of the whale has certainly not atrophied.

So whales and men diverged from the common ancestry, one to become the most lordly form of life in the oceans, and the other to become the dominant animal on the land. The day came when the two would meet. The meeting was not a peaceful one, in mutual recognition of each other's worth. As usual, it was man who set the terms—and he chose battle. It was a one-sided battle where man wielded the weapons, and the whales did the dying.

The bloody tale of men and whales has its beginnings in forgotten times when a few coast-dwelling tribes began putting to sea in skin boats or dugout canoes to test their hunters' skills against the monsters they first encountered as mountains of fat and meat when dead whales washed up on their shores.

In the northern hemisphere, such primitive people were hunting the Biscayen Right Whale, and probably the now extinct Atlantic Grey Whale, off the coasts of Portugal at least as early as 2000 B.C.

In North America, Eskimoan people of the Thule

culture hunted the Arctic Right Whale, while Indians on the Pacific Coast pursued the Grey Whale, and still other Indians on the Atlantic Coast took both Greys and Humpbacks.

In all cases the methods used were essentially the same. Paddlers in open boats tried to approach a whale close enough to let one of their number strike it with a barbed-bone or flint-headed harpoon, to which a rawhide line and a skin float were attached. Frequently the shallowly embedded weapon would break off or pull loose; or the boat would be swamped in the flurry as the whale sounded; or the line would part; or the float would be carried far beyond the range of the hunters to pursue it.

Rarely (and it must have been *very* rarely indeed), the hunters were able to stick with the whale, festooning it with more and more harpoons and floats until, eventually, it tired and they could pull alongside and try to kill it with thrusts from fragile lances. Since they could seldom hope to penetrate to a vital spot, they literally had to bleed it to death, a dangerous procedure during which the frenzied animal might not only crush their boats but might well crush them too. If they did succeed in killing the giant, they still had to tow it to the nearest beach, a task which, under adverse conditions of wind and tide, might take many hours or even days, or prove impossible.

Tribal traditions, together with the scarcity of whale bones in ancient kitchen middens, make it clear that any primitive whaling community which managed to kill two or three whales a year was doing rather well for itself. There was no need to kill more anyway. These people killed only to eat, and it took a long, long time for a handful of families to eat a whale. Consequently, early man posed no real threat to the continuing prosperity of the whale nation.

Nor was there much danger from modern man either until, during the 13th and 14th centuries, Europeans began building ships that could keep the sea. One of the earliest uses to which such ships were put was for pelagic, or open ocean, whaling; and, apparently, the first people to dare this chase were the Basques, who "fished" for the Biscayen Right and the Atlantic Grey Whale, not only because these were common in their waters but because they were slow-moving and rather unwary and, what was essential, did not sink when killed. With luck, a Basque ship could sail close enough to a Grey or a Right Whale so that the harpooner poised in the bows could strike into it with a heavy wrought-iron harpoon made fast to the ship by a strong warp, which not even a whale could easily part. The drag of the ship would eventually exhaust the animal and it could then be lanced to death with little risk.

The Basques still towed dead whales to shore for disposal but there had been a momentous change in the purpose for which they were being killed. These new whalers did not catch them for food. Instead they stripped off the layers of blubber and cut out the baleen plates. Then they turned the monumental carcasses adrift into the sea.

Only the oil and baleen were wanted now; the oil to fuel the lamps of an increasingly urbanized European society, and the baleen for the manufacture of "horn" windows and utensils. Thus the whale had been transformed from edible game into an article of commerce. When that happened man ceased to be a pin-prick irritant to the whale nation and became a deadly enemy. From this time forward whales were slaughtered without quarter and with every weapon and by every method the planet's most accomplished killers could devise.

The Basques were already efficient enough. By the end of the 15th century they had so reduced the Biscayen Right Whales that the species was hardly worth the hunting, and they had evidently exterminated the eastern population of Atlantic Grey Whales. However, far to the northward lay an even larger population —the Arctic Right, Bowhead or Greenland Whale, as it was variously called. In pursuit of this immensely abundant species (it is estimated there were more than half a million Arctic Right Whales before the great hunt began), the Basques had invaded Greenland waters by 1410 and were whaling off Labrador and Newfoundland by 1440. They still relied on shore stations established on the as yet officially "unknown" coasts to which they towed their catch for "cutting-in" and rendering the blubber. However, near the end of the 15th century, the Basques made another great stride forward. They invented and perfected ship-borne tryworks so whales could now be cut-in and rendered at sea.

From that day pelagic whaling exploded into a rapacious, world-wide slaughter of all those whales which were slow enough to be caught by sailing ships, and fat enough not to sink when killed. These were primarily Sperms, Humpbacks, Greys and Rights. By the mid-1800's there were as many as two thousand ships mercilessly sweeping the North and South Atlantic, both Pacifics, and the Indian Ocean, every year. They sailed from the New England States, Holland, the Baltic States, Norway, France, England, and a score of other places. They earned huge fortunes for the money-men at home, and by 1880 they had reduced to scattered remnants the once vast population of the great whale species which they pursued.

The slaughter had been so tremendous that, as the 19th century began to wane, it appeared that the hunt was coming to an end for want of whales . . . or, more accurately, for want of whales that men could catch.

There were still—and this was something which infuriated whalers and businessmen alike—enormous numbers of great whales in the sea. These were the baleen whales of the group called rorquals—Blues, Fins, Seis and a few lesser species. The rorquals included the largest, swiftest and undoubtedly the most intelligent of whales.

Quite apart from their wariness and the fact that most of them were capable of speeds of at least twenty knots, their relatively thin blubber layer failed to give them the positive buoyancy which had proved so fatal to the Greys, Humpbacks, Rights and Sperms. Consequently if, by exceptional good luck, a sailing ship managed to catch and kill one of the rorquals, the monster promptly sank, and that was that.

For a brief time it looked as if the rorquals would remain out of man's reach; but then the Norwegians, the most ruthless sea-marauders of all time, and by far the most accomplished killers of marine life, stepped in and took a hand. About 1860 they turned their hard blue eyes upon the rorquals and put their Viking minds to work. Within ten years they had found the means to doom, not only the rorquals, *but all surviving great whales in all the oceans of the earth.*

They attacked with three new weapons. First was the whale gun: a cannon which fired a heavy harpoon with a line attached deep into the whale's vitals, where a bomb exploded, ripping the animal apart internally and setting the broad barbs of the harpoon so they could not tear loose. The second was the steam catcher: a small, steam-powered vessel of great speed and manoeuverability which could match the rorquals' speed. The third was a hollow lance which was thrust deep into the dead whale and through which compressed air could be injected until the whale inflated and became buoyant. With these inventions the Norwegians took virtual control of world-wide whaling.

By the turn of the century their shore stations (for processing the carcasses) had spread like a pox along almost every coast in the world near which whales were found. In 1904 there were eighteen such factories on the shores of Newfoundland alone, processing an average of twelve hundred whales, most of them rorquals, every year!*

---

*Not all the hunting of rorquals was done with the newly devised harpoon gun. During the first decade of the 20th century Norwegians were killing Seis and Fins in a fiord near Bergen by a method so barbarous that it is hard to credit. The whales were driven into the long fiord by boats and the entrance was barred off with nets. The great animals were then speared with lances whose blades had been dipped in the rotting flesh of whales killed earlier. Infection set in and the trapped whales died horribly of septicemia or gangrene.

The world-wide slaughter was enormous and the profits even more so. By 1912 all the great whales, including Blues, both species of Rights, Fins, Sperms and Humpbacks, had nearly vanished from the North Atlantic and, with the addition of the Greys, from the North Pacific as well.*

It is likely that several of these species would have become extinct in the northern hemisphere had it not been for the outbreak of the First World War, which gave the surviving whales in northern waters a brief surcease, though not enough time to recover. The remnant survivors would have been quickly finished off if the Norwegians had returned heavily to the attack after the war was over.

That they did not do so was due to the discovery by the Norwegians about 1904 of an immense and hitherto untouched population of whales in the Antarctic Ocean. Here, during the decades that had almost emptied the other oceans of great whales, a sanctuary had existed. When the Norwegians nosed it out, fleets of swift, merciless catchers swarmed southward to begin a new and even more thorough butchery of the whale nation from shore bases in the Falkland Islands and South Georgia.

Then, in 1922, a Norwegian named Carl Anton Larsen, whose name deserves to be forever remembered in equal opprobrium with that of Sven Foyn, inventor of the harpoon gun, brought about the ultimate refinement in commercial whaling. He invented the modern factory ship. In its essential form, this is a very large cargo vessel with a gaping hole in her stern through which a hundred-ton whale can be hauled up into a combined floating abattoir and processing plant. With her coming, away went the pressing need for shore stations and the long, time-wasting tows to land. Accompanied by fleets of catchers, buoy boats and tow boats, and stored for a voyage of six months or

---

*Grey Whales were finally exterminated in Atlantic waters sometime toward the end of the 18th century.

more, the factory ships could penetrate far southward to the edge of the Antarctic ice itself and could range the whole expanse of the Antarctic seas.

The ensuing slaughter of an animal population is unparalleled in human history. The combination of man's genius for destruction together with the satanic powers of his technology dyed the cold, green waters of the Antarctic crimson with the heart's blood of the whale nation. The massacre built to a crescendo in the early 1930's when as many as 80,000 great whales died each year!

The outbreak of the Second World War brought a pause to the purposeful slaughter of whales in the Antarctic, but it brought new setbacks to the slowly recovering populations of whales in the other oceans of the world. The war at sea was primarily a war between submarines and surface ships, and the submarine—which is no more than a manmade imitation of a whale, in form—came under increasingly sophisticated and sustained attack as the war went on.

Such technological marvels as sonar and Asdic were refined to detect and follow underwater objects with great accuracy, and could guide depth charges, bombs, and other deadly devices to the unseen target. Although, to my knowledge, the matter has never been investigated or even publicly discussed, there is no doubt that tens of thousands of whales were killed by the men who hunted submarines with ships or planes.

A commander in the Royal Canadian Navy who served four years in corvettes, frigates and destroyers in the North Atlantic told me he believed a high percentage of the depth-charges fired from his ships had been directed at submerged whales rather than submarines. The drifting carcasses of bombed or depth-charged whales were a common enough sight to lookouts aboard naval and merchant ships. Wars are deadly, not only to mankind, but to those most innocent bystanders, the other forms of life which share the planet with us.

Here is as good a place as any to reply to a question I have sometimes been asked. Why is it that, if whales have such large and well-developed brains, they have not been able to avoid destruction at man's hands? The answer seems obvious. The whales never dabbled in the arcane arts of technology and so had no defence against that most deadly plague. In time they might have evolved a defence, but we gave them no time. The answer raises a counter question: Why is it, if man has such a remarkable intelligence, *he* has been unable to avoid an almost continuous acceleration of the processes of self-destruction? Why, if he *is* the most advanced of beings, has he become a threat to the survival of all life on earth?

At the end of the Second World War, despite the fact that the Antarctic whale population had shown no increase, the whalers went back to work with renewed energy and with even deadlier weapons. Sophisticated sonar gear, radar, and spotting aircraft operating from immense new floating factories (some as big as 30,000 tons), were combined with powerful new catchers that could make twenty knots with ease. This combination ensured that any whale which came within the wide-reaching electronic ken of the killer fleet stood no more than a fractional chance of survival.

By the early 1950's the Blue Whale was rapidly approaching "commercial" extinction so the hunters turned their main effort on the Fins. They were so successful that, by 1956, there were no more than 100,000 Finners left alive out of a population estimated to have been nearly a million at the turn of the century. And in 1956, 25,289 Finners—one quarter of those remaining in the seas—were slaughtered! By 1960 there *may* have been 2500 Blue Whales left in all the oceans of the world (of whom less than a thousand now survive), and perhaps 40,000 Fins in the Antarctic. They were so few, and so widely dispersed, that it hardly paid the pelagic whaling fleets to hunt them anymore, and so they turned to the lesser rorquals. They began exterminating the Seis.

Although the official whaling returns of the late 1950's made it obvious that the great whales were entering their final hours on this earth, nobody took any effective action to halt the butchery. When a few worried biologists suggested that the whaling industry should establish meaningful quotas which would result in the whalers being able to make a sustained harvest indefinitely, while allowing the whale nation to at least partially recover its numbers, they were ignored. The owners of the whaling fleets made it patently clear that they were determined to hunt the whales to extinction, and the devil take the hindmost.

Hardly a voice was raised in public against this calculated policy of extermination. On the contrary, there was a spate of novels, non-fiction books and motion pictures, many of which seemed to glorify the slaughter, and all of which praised the hardihood and manliness of the whale killers.

It is true that in 1946 an organization had been formed with the publicly stated intention of giving protection to the threatened species of whales and of regulating the hunt. This was the International Whaling Commission whose headquarters were (and remain) in Norway, which also happened to be headquarters for the world's most efficient whale killers. But despite the employment of many good and dedicated men, the Commission was run for, and by, the whalers; and in such a manner that, instead of helping to preserve and conserve the vanishing whale stocks, it served as a cynical device to divert attention from the truth. It served to mask the insatiable greed which lay behind the slaughter, by promulgating regulations which appeared wise and humane but which, in fact, were useless . . . and sometimes worse than that.

One of the first actions of the Commission was to institute a quota system whereby each nation was allowed to kill only so many whales. It was a totally meaningless gesture since the quotas were set, and have since been consistently maintained, at levels far higher than the whale populations could support. There were regulations against the taking of undersize whales, or of cows accompanied by their calves; and these were honoured mostly in the breach. But the most hypocritical of all were the regulations, declared with great fanfare, which ultimately prohibited the killing of Blues, Humpbacks and all species of Right Whales. These were brought into effect *only after* all these species had been brought so close to extinction that they were no longer of any major commercial value and were, in fact, all threatened with *biological* extinction. These regulations were promulgated . . . but they were not enforced! The Japanese, for example, evaded them by pretending to discover a new species of whale in the Antarctic. They named it the *Pygmy* Blue Whale and, since it was outside the quota and not included in the prohibition on killing protected species, they busily cleaned up this last viable pocket which might very well have provided a nucleus for the return of the doomed Blue Whale. Furthermore, almost all pelagic whaling fleets, of whatever nationality, took protected whales upon occasion, giving the excuse that these were cases of mistaken identity. Still worse, many nations allowed their whalers to kill significant numbers of protected species for "scientific research." Nearly 500 Greys (also a protected species) out of a world population of under 10,000 were killed under scientific permits between 1953 and 1969 by Russian, Canadian and United States whalers, with the Americans alone taking 316 of these. During the past three years whalers on Canada's east coast have killed 43 rare Humpbacks (the survivors of this once numerous species now number fewer than 2000) in the name of science. While it is true that scientists examined most of these sacrificial whales, adding presumably to their anatomical knowledge of the dead beasts, it is also true that the carcasses became the property of the whaling companies, who processed them commercially, for profit.

Because of the inestimable damage it has done by assuming the guise of champion to the beleaguered whales, and so gaining acceptance in the public eye as *the* authority on the subject, I must emphasize that the International Whaling Commission has served only to hasten the doom of most of the species it has pre-

tended to protect, while concealing the magnitude of
the crime against life which has been, and *is being,*
committed by the whalers on behalf of powerful individuals, industries, and governments; and, in the last
and inescapable analysis, on behalf of all of us.

Chapter 5

It was during the winter of 1911 that Uncle Art saw the disappearance of the great whales from the Sou'-west Coast.

"Back about 1903 they Norway fellows built a factory in a cove eastward of Cape La Hune. Called it Baleena and, me son, 'twas some dirty place! They had three or four steam catching boats with harpoon guns, and they was never idle. Most days each of them would tow in a couple of sulphur bottoms or Finners* and the shoremen would cut 'em up some quick. No trouble to smell that place ten miles away *up*wind!

"And floating whales! When they got the blubber stripped off they turned the carcasses loose, the meat all black, and they blasted up so high they floated nigh out of the water. Some days when I been offshore I t'ought a whole new set of islands had lifted overnight. Five or ten in sight at once, and each one with t'ousands of gulls hangin' over it like a cloud.

" 'Twas a hard winter for weather and I never got to the Penguins as much as in a good year, but whenever I was there I hardly see a whale. Then, come February and the weather got fittin', I made a run for the grounds. Was a good sign of fish so I stayed on the islands six or seven days. One morning 'twas right

_____
*Colloquial names for Blue Whales and Fin Whales.

frosty but nary a pick of wind. I was workin' a trawl near the Offer Rock when I heard this girt big sound. It kind of shivered the dory.

"I turned me head and there was the biggest Finner I ever see. My *son!* He looked nigh as big as the coastal boat. He was right on the top of the water and blowing hard, and every time he blowed the blood went twenty feet into the air. He stayed where he was to, a dozen dory lengths from I, and I could see there was a hole blowed into his back big enough to drop a puncheon into. One of them bomb harpoons must have took him, and then the line parted.

"Now I got to say I was a mite feared. There's no tellin' what ary wounded beast will do. I was tryin' to slip me oars through the t'ole pins, quiet like, when he began to move straight for I. Was nothin' to be done but grab the oar to fend him off, but he never come that close. He hauled off to starboard, and then he sounded, and I never saw he again . . . no, nor any of his like, for fifty year."

Until after the Second World War there were almost no sightings of great whales off the south coast of Newfoundland. Then, in the late 1940's, U.S. Naval aircraft flying out of the leased base at Argentia in southeast Newfoundland began spotting an occasional big whale. News of these sightings came to light in the mid-1950's when it was learned that whales had become a useful addition to the Navy's anti-submarine training. Aircraft crews engaged in practice patrol work had been instructed to pretend that any whales they spotted were Russian submarines. The whales became targets for cannon fire, rockets, bombs and depth charges!

In 1957 an outcry by Harold Horwood, a crusading columnist on the *St. John's Evening Telegram,* resulted in a promise from the Argentia officials that whales would no longer be used as targets. However, the number which had been attacked, wounded or killed over a ten-year period was never released. Presumably it was classified information.

The Americans were certainly not the only military people who were sacrificing whales "in defence of freedom." Most countries with sea coasts to guard, and aircraft and ships to exercise, were probably also abusing whales in like manner, and may very well still be doing so.

Despite their reception by the u.s. Navy, a few great whales continued to filter into the vacuum created in south Newfoundland waters by the Norwegians half a century earlier. These may have been fugitives from the coastal waters of northern Newfoundland and southern Labrador where, in 1945, the Norwegians returned to the attack and established shore whaling stations at Williamsport and Hawkes Harbour. The Finners in this region had managed a remarkable come-back during thirty years of relative freedom from the hunters. It was short-lived. In the six years up to 1951, these two stations between them killed 3721 Fin Whales before they began to run out of targets and both stations became "uneconomical" and were closed.

Wherever they came from, little pods of Fins and Seis, a few Humpbacks and Sperms, and even the odd Blue Whale, were seen with increasing frequency during the late 1950's in the seas south of Newfoundland and east of Nova Scotia. In December of 1961, Uncle Art and Uncle Job were hauling herring nets near Hunts Island in the Burgeo archipelago when two Fins surfaced close by.

" 'Twas the finest sight I t'inks I ever had, to see they whales come back!" Uncle Art recalled.

"They had no fear of we. The herring was thick as smoke that month and they whales was hungry as ary wolf. They made a sweep or two close to our nets and the herring rose right out of the water ahead of she. They drove into one net and filled it up so full we had to go ashore and get some extra hands to help us haul it up."

These two whales were joined by another adult Fin and the little group remained in Burgeo waters until April 1961, when they and the herring disappeared together.

Their presence at first caused some concern among the younger fishermen who knew nothing about whales except by hearsay. They were afraid for their fishing gear, particularly because of an incident which had taken place the previous summer. A 20-foot Greenland Shark had invaded the narrow waters between the islands and blundered its way westward, leaving a trail of shredded salmon nets behind, until the weight of twine and anchors it was hauling grew too much for even its mighty muscles and it drowned.

The shark had been a huge enough and destructive enough monster but he was dwarfed by the Fin Whales who had the potential to wreak much greater havoc. They did nothing of the sort. Not a net or a mooring was touched nor a fishing boat threatened. After a few weeks the fishermen came to take the whales as much for granted as the whales apparently took them. To my knowledge, there was never any suggestion made by fishermen that "something should be done" about the whales. Perhaps their acceptance of the great animals was partly due to the fact that they had no weapons which could have been effective against such giants, but I think not. Their reaction was the normal one of hunting man in a natural environment toward another form of life which he cannot make use of, does not need, and which poses no threat or inconvenience to him. Live and let live was the attitude of the Burgeo fishermen.

Unfortunately, not all Newfoundlanders felt this way. Word of the return of the great whales soon reached the sharp ears of Premier Smallwood who was not the man to let an opportunity for the exploitation of a natural resource pass him by.

Smallwood had already tried to profit by the presence in Newfoundland waters of a small but abundant toothed whale called the Pothead. Potheads seldom grow more than eighteen feet in length but they live in large herds sometimes numbering in the hundreds. Their favourite food is small squid, in pursuit of which they often come deep into Newfoundland's narrow fiords. Here they are very vulnerable to attack.

Smallwood's plan was to establish a mink industry and to feed the mink on the meat of Pothead Whales. In quick order several mink ranches were constructed (one of the largest of them belonging to Premier Smallwood himself), and the slaughter of the Potheads began. Dildo was the favourite killing ground. When a herd entered this fiord, some boats would bar off the entrance, while others, whose crews were armed with shotguns and other noise makers, panicked the little whales and drove them aground in the shallows at the head of the fiord. By 1965 more than 50,000 had been axed, clubbed and stabbed to death in Newfoundland waters for mink food, so decimating the Pothead stock that three Norwegian catchers had to be hired to hunt thirty-foot Minke Whales, the smallest members of the baleen family, which had been ignored until then but which are now coming under heavy attack.

Smallwood also had designs on the great whales. He invited foreign interests to again revive the slaughter off the northern coasts of Newfoundland, offering incentives which amounted to guarantees that the whalers could do almost as they pleased, plus subsidies on a grand scale. The Japanese were all too happy to accept and they re-opened the factory at Williamsport while a consortium of Japanese and Norwegians expanded the Dildo operation to include the killing of the big rorquals. Soon the catchers from these two stations were ranging far at sea, into the Straits of Belle Isle and up the Labrador Coast, and taking between three and four hundred Finners a year. Both stations are still in operation, having killed 2114 Fin Whales (together with several hundred Seis, Sperms and Minkes) between 1965 and 1971. Under a quota system authorized by the Canadian government, these stations, plus a third in Nova Scotia, will be allowed to kill 360 Finners in 1972 . . . which is about the limit of their maximum catching effort . . . and this from a rapidly declining population of western North Atlantic Fin Whales which, according to at least one biologist, may already have plunged to below 3000 individuals. It is likely that the three stations will simply not be able to find and kill 360 Finners but, no matter. There are *no quotas* on how many whales of other species they may slaughter and so there is little likelihood that they will be deprived of a good margin of profit for their year's work. However, if they do manage to fill the generous Fin Whale quota, they, and the Canadian government, will have made a significant contribution to the ultimate destruction of a species whose worldwide numbers are now thought to be less than 60,000.

The tranquil acceptance of the Fin Whales at Burgeo was in sharp contrast to an incident I witnessed at about this time at St. Pierre, the capital and only port for the French islands of St. Pierre-Miquelon which lie a few miles off the south coast of Newfoundland. Most of the inhabitants there are fishermen too, but St. Pierre itself is full of shops, tourist establishments, ship repair facilities, and people whose loyalties lie with the modern industrial society.

On a moonless night in August 1961, my schooner lay moored to a rotting dock in St. Pierre harbour. About midnight I went on deck to smoke a pipe and enjoy the silence; but the quiet was soon broken by what sounded like a gust of heavy breathing in the waters almost alongside. Startled, I grabbed a flashlight and played its beam over the dark waters. The calm surface was mysteriously roiled in great, spreading rings. As I puzzled over the meaning of this phenomenon, there came another burst of heavy exhalations. I swung the light to port and was in time to see one, three, then a dozen broad black backs smoothly break the oily surface, blow, then slip away into the depths again.

I was seeing a school of Potheads who had made their way into the sewage-laden waters of the inner harbour. They must have had a pressing reason, for no free-swimming animal in its right mind would have entered that cesspool willingly. The skipper of a local dragger later told me he had met a small group of Killer Whales close to the harbour channel on the day the Potheads entered. Killer Whales have been given a ferocious reputation by men; one not at all deserved, but it is true that they will occasionally make a meal of a Pothead calf, and the Potheads in St. Pierre harbour were accompanied by several calves.

When I went to bed, the whales were still circling leisurely. I slept late, to be awakened by the snarl of outboard engines, by excited shouting, and by the sound of feet pounding on my deck. When I thrust my head out of the hatch, I found what appeared to be about half the male population of St. Pierre, accompanied by a good many women and children, closely clustered along the waterfront.

There was a slight fog lying over the harbour. In and out of it wove two over-powered launches, roaring along at full throttle. In the bow of one stood a young man wielding a homemade lance which he had made by lashing a hunting knife to the end of an oar. In the second boat was another young man, balancing a rifle across his knees. Both boats were in furious pursuit of the Potheads which numbered some fifteen adults and six or seven calves.

The whales were very frightened. The moment one of them surfaced, the boats tore down upon it, while gunners on the shore poured out a fusillade of shots. The big animals had no time to properly ventilate their lungs but were forced to submerge after snatching a single breath. The calves, choking for oxygen, were often slow in diving. Time after time the harpooner got close enough to ram his hunting knife into the back of one of them so that long streamers of crimson began to appear on the filthy surface of the harbour. It was obvious that neither the gunfire—mostly from .22 calibre rifle—nor the lance were capable of killing the whales outright; but it did not appear that killing them was the object. In truth, what I was watching was a sporting event.

I was appalled and infuriated, but there seemed

to be nothing I could do to end this exhibition of wanton bloodlust. A fisherman friend of mine, Theophille Detcheverey, came aboard and I poured out my distress to him. He shrugged.

"That one in the big speedboat, he is the son of the biggest merchant here. The other, with the spear, he is from France. He came here two years ago to start a raft voyage across the Atlantic. But he don't get out of the bars until today, I think. They are pigs, eh? But we are not all pigs. You see, there is no fisherman helping them with their dirty work."

This was true enough, if of small comfort to the whales. The fishermen of St. Pierre had left for the cod grounds at dawn. When they returned in their laden dories late in the afternoon, the excitement in the harbour had reached a crescendo. All the fast pleasure craft available had joined in the game. The onlookers crowding around the harbour had become so densely packed it was hard to push one's way through. I had chased scores of them off my decks where they sought a better vantage point; and they had responded to my anger with derision. For ten hours, relays of boats had chased the whales. Clusters of men with rifles stood at the pierhead at the harbour entrance and every time the Potheads tried to escape in that direction they were met with a barrage of bullets which now included heavy-calibre slugs. Unable to run that gauntlet, the whales were forced to give up their attempts to escape in the only direction open to them.

Toward evening the whales, most of them now bleeding profusely, had become so exhausted they began to crowd up into the dangerously shoal water at the head of the harbour where the boats could not follow. Here they lay, gasping and rolling, until they had recovered enough strength to return to deeper water. Many times they swam directly under my boat, and they were beautiful . . . superb masters of the seas, now at the mercy of the bifurcated killer of the land.

At dusk the sportsmen called it a day and went

home to dinner. The audience departed. The fog rolled in thickly and silence returned. Again I sat on deck, and again the strange sibilant breathing of the whales kept me company. I could not go to my berth, knowing what must await them with the dawn. Finally I untied my little dinghy and rowed out into the darkness of the fog shroud. I had a vague hope that I might be able to drive the herd out of the harbour before daylight brought a renewal of their ordeal.

It was an uncanny experience, and a nerve-wracking one, to row my little cockleshell silently through that dense and dripping fog, not knowing where the whales might be. The size of them—the largest must have been nearly twenty feet long—and their mysterious and unseen presence intimidated me. I felt extraordinarily vulnerable, detached from my own world, adrift on the lip of a world which was utterly alien. I thought, as a man would think, that if there was the capacity for vengeance in these beasts, surely I would experience it.

Then, with heart-stopping suddenness, the entire pod surfaced all around me. A calf blew directly under one upraised oar and my little boat rocked lightly in its wash. It should have been a terrifying moment, but it was not. Inexplicably, I was no longer afraid. I began talking to the beasts in a quiet way, warning them that they must leave. They stayed at or near the surface, swimming very slowly—perhaps still exhausted—and I had no difficulty staying with them. Time after time they surfaced all around me and although any one of them, even the smallest calf, could have easily overturned the dinghy, they avoided touching it. I began to experience an indescribable sense of empathy with them . . . and a mounting frustration. How could I help them to escape from what the morrow held?

We slowly circled the harbour—this strange flotilla of man and whales—but they would not go near the harbour mouth, either because they knew the Killer Whales were still in the vicinity or because of the vicious barrage of bullets with which men had greeted their every attempt to escape during the daylight hours.

Eventually I decided to try desperate measures. At the closest point to the harbour entrance to which they would go, I suddenly began howling at them and wildly flailing my oars against the water. Instantly they sounded, diving deep and long. I heard them blow once more at the far side of the harbour but they never came close to me again. I had done the wrong thing— the human thing—and my action had brought an end to their acceptance of me.

The whales were still in the harbour when dawn broke. During the long evening in the bars, the ingenious sportsmen of St. Pierre had set the stage for a massacre.

Early in the morning, just as the tide was beginning to ebb, half a dozen boats came out and formed a line abreast at the harbour mouth. Slowly, they began to sweep the harbour, driving the herd closer and closer to the shoals. When the whales sounded and doubled back, they were again met with rifle fire from the breakwater as on the day before. One of the largest beasts seemed to be leading these attempts to escape, with the rest following close in its wake. It looked like a stalemate until three small whales became momentarily separated from the pod as it came under the fusillade from the breakwater. They gave way to panic. Fleeing at full speed on the surface, and close-harried by a fast speedboat, they torpedoed across the harbour and into the shoals where the tide was dropping fast. Within minutes they were hopelessly aground.

Howling like the veriest banshees, men and boys armed with axes and carving knives leapt into the knee-deep shallows. Blood began to swirl thickly about them. The apparent leader of the pod, responding to what impulse I shall never know, charged toward the three stranded and mutilated whales. There was a wild melée of running, falling, yelling people; then the big whale stranded too. The rest of the herd, following close behind, were soon ashore as well. Only one calf remained afloat. It swam aimlessly back and forth just beyond the fatal shoals, and for a few minutes was ignored as the boats crowded in upon the herd and men leapt overboard, jostling one another in their lust to have a hand in the slaughter. Blood from one impaled whale spouted high over their heads—a red and drenching rain. Men flung up their ensanguined faces, wiped the blood away, and laughed and shouted in the delirium of dealing death.

Finally someone noticed the calf. Arms, red and savage, pointed urgently. A man leapt into his speedboat. The engine roared. He circled once at top speed then bore straight at the calf which was in such shoal water it could not sound. The boat almost ran up on its back. The calf swerved frantically, beat its flukes wildly, and was aground.

The slashing and the hacking on that bloody foreshore continued long after all the whales had bled to death. A crowd of four or five hundred people drank in the spectacle with eager appetite. It was a great fiesta in St. Pierre. Throughout the remainder of the day there was a crowd standing and staring at the monstrous corpses. I particularly remember a small boy, who could not have been more than eight years of age, straddling a dead calf and repeatedly striking into its flesh with a pocket knife, while his father stood by and encouraged him.

Nor were the "townies" of St. Pierre the only ones to enjoy the spectacle. Many American and Canadian tourists had witnessed the show and now were busy taking pictures of one another posing beside the dead behemoths. Something to show the folks back home.

It was a grand exhibition . . . but the aftermath was not so grand. Those many tons of putrefying flesh could not be left lying where they were. So, on the following day, several big trucks appeared at the shore where lay the carcasses of twenty-three Pothead Whales. One by one the whales were hauled up by a mobile derrick and either loaded aboard the trucks or, if they were too big, chained behind. Then the trucks carried and dragged the bodies across the island to a cliff where, one by one, they were rolled over the steep slopes . . . and returned to the freedom of the seas.

Chapter 6

A curiosity about the whale nation has been a part of me for as long as I can remember. When I was a very small child my grandfather used to sing me a song that began:

In the North Sea lived a whale . . .
Big of bone and large of tail . . .

The song went on to describe how this particular whale was the master of his world until the day when he espied a stranger in his domain: a big, gleaming silver fish who stubbornly refused to acknowledge the whale's mastery. The whale grew angry and slapped the interloper with his tail. That was a fatal error, because the strange fish was actually a torpedo.

The moral of the song (all children's songs of that era had a moral) must have been that it does not pay to be a bully. I never understood it that way. The song haunted me because my sympathy was entirely with the whale—the victim, so it seemed to me, of a very dirty trick.

As I grew older and became more and more fascinated by non-human forms of life, the whale became a symbol of the ultimate secrets which have not yet been revealed to us by the "other" animals. Whenever anything came to hand about whales, I read it avidly; but the only thing which seemed to emerge with certainty from all my reading was that the whales appeared to be doomed by human greed to disappear and to carry their secrets with them into oblivion.

Before coming to live in Burgeo, I had never actually seen one of the great whales. Knowing how rapidly they were being destroyed, I never really expected to see one. However, I had not been there long when I heard about the little pod of Finners which had spent the previous winter among the Burgeo Islands. The possibility that they might come back again was intensely exciting and contributed something to my decision to make our home in Messers Cove.

Shortly after I first met Uncle Art, I asked him if he thought the whales would return. He assured me they would, but I hardly dared believe him until a day just before Christmas in 1962.

It was a cold and pallid day with a hazed sky and sun dogs circling a half-veiled sun. Claire and I were in our kitchen, reading, when Onie Stickland came quietly through the door to tell us that whales were spouting just off Messers Head.

Seizing binoculars, we followed him out along the snow-crusted promontory. Not more than a quarter of a mile off the headland, several quick, high puffs of vapour bloomed and hung briefly in the still air. We could catch only elusive glimpses of the great beasts themselves: slick black mounds, like moving rocks, awash in the jet-dark waters. It was enough. For me it was a moment of supreme excitement and supreme awareness. The secret was here—was now—was on my own doorstep.

Four Fins comprised the family that spent the rest of that winter in Burgeo waters, and it was a bad day indeed when we could not spot them from our seaward windows. Uncle Art was so delighted to have them about that, I truly believe, he set most of his herring nets, not to catch herring, but simply to justify the hours he spent on the water watching his gigantic friends. Ashore, he was equally happy to sit for hours telling me what he had learned during his long life about leviathan; and slowly, slowly, I began to steal some glimpses through the shroud of mystery.

During the succeeding years, the whales returned to our coast early each December and remained with us until the herring departed, usually sometime in April. Each winter we looked forward with unabated eagerness to their arrival. We were not alone in our interest. In general, the fishermen of Burgeo seemed to hold the great beasts in a kind of rough and friendly regard. There was no conflict between whales and men. The whales remained scrupulous in keeping clear of the fishing gear; nor could they be looked upon as competitors for the herring since our fishermen were only interested in taking small quantities for use as trawl bait, plus a few tubs to be salted down for table use. Since the herring were present in millions, if not billions, there were far more than enough of them for whales and men together.

The whales and the inshore fishermen developed a remarkable familiarity. Whales would often surface only yards from a dory, a skiff, or even a forty-foot longliner; blow, draw in a huge draft of air, then return unconcernedly to their fishing while the human fishermen went on as unconcernedly with theirs.

It was Uncle Art's conviction that the whales looked upon our fishermen with an almost benevolent tolerance, as those who are past masters of a complex trade may sometimes look upon willing but not very bright apprentices. This is how matters stood until the arrival of the "foreign" purse seiners on our coast.

Although anatomists and other such can tell us something about the mechanisms of dead whales in terms of what they are, rather than of what they do, science remains surprisingly ignorant about the activities and behaviour of *living* whales and understands even less about their special capabilities in their aquatic world. A friend of mine, who is one of the foremost cetologists of our time, recently summed up the state of our knowledge in these words: "The little we biologists know about whales in life would hardly provide enough material for an essay by a high school student."

In view of the fact that man's interest in the great whales, through the millenniums, has been largely restricted to bringing them to death, this is not very surprising. It has only been within the last few decades that we modern technological men have made any real effort to study them as living creatures; and by the time our scientists began to show some interest in the matter, there was only a remnant population of many species left and that, however, so widely dispersed over such an immense realm of waters that we landsmen could count ourselves lucky to catch even occasional glimpses of them. A modern scientist attempting to plumb the secrets of whale life is in much the same predicament as the denizen of another planet would be if, suspended high above the atmosphere, he tried to comprehend the intricacies of human life through the cloudy ocean of air surrounding us. Fortunately, we do not have to rely solely upon what professional science has been able to piece together.

Because Fin Whales are herring eaters (at least in certain seasons and in certain parts of their vast oceanic range) and because the herring strike inshore every winter along the ice-free southern coast of Newfoundland, there is a period when the great mammals live almost at the portals of our world. And because of men like Uncle Art, who are possessed of that abiding curiosity about other forms of life which is the hallmark of natural man, we actually know rather more about the great whales than my scientist friend concedes.

It is proper for me to acknowledge my debt to such men as Uncle Art, and to warn the reader, if such a warning is needed, that in almost all that I have to tell about the Fin Whale, I have drawn heavily on the observations and on the intuition of such natural observers. For instance, I have been able to find nothing in writing to explain even such an apparently basic thing as how a Fin Whale gets his supper. It was Uncle

Art who gave me the key to this minor secret. My own subsequent observations enlarged upon the picture and, when combined with some of science's discoveries about whale sonar, enabled me to come to a broad understanding—a somewhat awestruck one—of how a Finner feeds himself.

When Uncle Art and I stood on Messers Head on the winter day in 1967 watching that marvel of technology —a modern herring seiner, competing with a family of Fin Whales—Uncle Art was moved to amusement. He had reason to be.

Consider one of those new seiners: a hundred-odd feet of steel, diesel power and complex electronic gear, all designed, with the ultimate in modern man's technical skill, to pursue and catch a little fish about a foot in length.

The seiner begins its work by first locating a herring school with a sophisticated echo-location system whereby a pulse of sound is transmitted through the water to be reflected back by any object it encounters. The returning echoes "read off" on a slowly unrolling strip of sensitized paper, and a dense school of herring will show clearly on this record. Depth and distance are also indicated. The seiner follows the electric scent and closes with the school. When the ship is in position she "shoots her seine," a fine-mesh net which is laid in such a fashion as to encircle the school. The net is then "pursed": its bottom edges being drawn together and its circumference diminished until the catch is concentrated, as in a huge dip net, alongside the ship. A big suction pipe is then lowered into the purse and the herring are pumped from the net and spewed into the vessel's holds. From beginning to end it is a complex operation which may take several hours to complete.

Now consider a big Fin Whale: as much as eighty feet of bone, muscle, sinew and brain, which has evolved into a living instrument of maximum efficiency for the pursuit of a little fish about a foot in length.

The whale locates a herring school either by sight (the Finner's underwater vision is excellent) or by means of a highly sophisticated echo-location system of his own, transmitting a pulse of very low-frequency sound through the water.* This low-frequency "sonar" has a broad beam effect. If there is anything in front, below or above him within a range of one to several miles, he will "see" it in the pattern of returning echoes. If it is a herring school—and the whale can presumably distinguish and identify herring as distinct from other species of fish of similar size—the whale heads for the target at his normal underwater cruising speed, which seems to be about eight knots.

As he closes with the school he accelerates in a burst of speed which may reach twenty knots. Closing with his target, he alters course and at torpedo velocity begins circling the school, spiraling steadily inward. As he does so, he turns on his side with his belly presented toward the herring. Unlike his grey-black back, his belly offers a huge expanse of glistening white with a very high light-reflecting value. Encircled by this flashing ring of light,† the herring jam tightly in upon one another in much the same manner as if they were surrounded by a seine.

When the circle (the "net" of reflected light) is tight enough and the herring sufficiently concentrated, the whale abruptly charges straight into the mob of little fishes with his enormous mouth agape.

Sheer speed, even when combined with his big mouth, might not alone suffice to bring him an ade-

---

*New evidence suggests that Finners may also have, and use, high-frequency, narrow beam sonar similar to that possessed by many of the toothed whales, capable of scanning objects at great distances, and perhaps also used for long-range communication.

†The Fin Whale's use of reflected light to "herd" herring explains another minor puzzle—his strikingly asymmetrical colour pattern. The white of the belly extends far higher up the flank on the whale's right side than it does on his left. I conclude that this is because the fishing whale circles a herring school clockwise, which means that his right side is always presented to the herring so that the barrier effect produced by the reflected light comes into play even before he turns on his side and begins to tighten the circle.

quate return for his effort; so he brings into play another and very special device. The whole underpart of his body, from directly beneath his mighty chin and extending aft to a point near his navel, is slit and pleated like a gigantic accordion. All rorquals have these slits, which in the Fin number about a hundred. When a Finner is travelling at speed and suddenly opens his mouth, the immense pressure of static water exerts itself on the whole gaping forefront of his body which promptly inflates to gargantuan size as the accordion pleats open to their fullest extent. Thus, instead of being able to engulf only a few barrels of salt water, and whatever herring are contained therein, he almost instantaneously ingests many tons of water together with its contents. Smartly now he closes his mouth, contracts the muscles controlling the pleats, and squirts the unwanted water out of apertures set at the corners of his jaws. The herring are tapped against the sieve of baleen plates. When there is nothing left in his mouth but herring, he uses his tongue to sluice them through his surprisingly small gullet into the first of his several holds, or stomachs. The entire operation takes about ten minutes.

Human seiners can only take herring, capelin and other such small fry when these are at or near the surface of the sea; but the Fin knows no such limitations and can probably go as deep as, or deeper than, any of the food fishes he pursues. Fin Whales can certainly cruise contentedly at a depth of at least 300 feet for as long as 30 minutes and on occasion can go much deeper and remain even longer. We do not know *how* deep they can go but one Fin which was harpooned with a device containing a depth recorder descended 1164 feet.

The fishing procedure I have described appears to be standard for daylight fishing; but what do Fins do at night? That they *do* sometimes fish in darkness is an observed fact. Echo location would work as well by night as by day, but the light-barrier effect presumably would cease to be of much value. Apparently the Fin does not employ the circle-and-concentrate system on very dark nights, relying instead on a straightforward rush executed at high speed against a school of prey fishes. This is probably a relatively inefficient procedure, and it may be that the only Fins who regularly hunt at night are pregnant or nursing females who need a heavier than normal, and more nearly continuous, food supply.

The similarities between the fishing methods of whales and those of technological man are fascinating, but there is an essential difference between them. After the whale has scooped up a ton or two of food, he ceases fishing and is free to do whatever it is whales choose to do with their spare time. The human herring seiner, on the other hand, operates on a different principle. Until the seiner is loaded to her marks with as much as 200 tons of herring, the human crew keeps at the killing.

Before modern man began his murderous exploitation of the seas, the oceans swarmed with herring . . . and with whales. Now that is changed. Having come close to eliminating the great whales, man is now rapidly doing the same thing to the herring. Today the once-famous herring fishery in the North Sea is fast declining. There are almost no herring stocks of any significant commercial value left off the coasts of Norway, England, or other European nations. A few years ago the British Columbia herring fishery, which was one of the richest in the world, was virtually fished out. By 1967, more than fifty big, modern British Columbia seiners had made the long voyage via the Panama Canal to Canada's east coast where they are now helping to clean the herring out of the Bay of Fundy, the Gulf of St. Lawrence and the waters adjacent to the southern coast of Newfoundland.

This fishery was extremely successful in its first few years of operation. In 1969 more than 120,000 tons of herring were taken by seiners in Newfoundland waters alone . . . and almost all of this mighty harvest of the sea was reduced to fish meal and fish oil for agricultural and industrial use. Only an insignificant part was used directly as food for man. However, by

1970, despite an even greater catching effort, the take had fallen by one-third. Samples taken in 1971 by fisheries biologists showed that the new-year classes (new generations) of herring were not coming along at anything like a sufficient rate to fill the gaps in the herring ranks. It is a competent prediction that by 1980 the herring fishery in eastern Canadian waters, if not in the entire North Atlantic Ocean, will have ceased because there will be almost no herring left.

Their disappearance will bring the threat of starvation to a vast array of commercially important fishes such as cod, halibut, haddock and even salmon, all of which depend heavily upon herring and upon which, in turn, the wildly proliferating human species is itself becoming increasingly dependent for *its* survival.

This is not something which worries the owners of the efficient new herring-reduction plants which have sprung up all along the Canadian Atlantic coasts. These plants (there is now one in Burgeo and it is due to triple its capacity before 1973) are highly-mechanized, low-cost, low-employment operations. Most of those built on the Sou'west Coast amortized their investment costs *in the first three years* of operation. They are now intensely profitable and the owners expect to continue making a big profit, even after the herring have been destroyed, by switching the attack to the other primary food of those large northern food fishes which man has used to sustain himself for countless centuries. The seiners will go after the capelin: a slim, beautiful little fish which is found only in northern waters but which may be as numerous at present as the herring were in the days before the purse seine fishery began.

The Norwegians have already begun seining capelin in eastern Atlantic waters in order to feed their reduction plants. Two Newfoundland plants are already experimenting with them. One of the executives of an international fisheries corporation happily told me he expects the capelin supply will "hold up for as long as five or ten years." What happens after that is something for which the fishing industry, apparently, has no great concern.

It will not be a happy prospect for the North Atlantic Fin Whales, if, indeed, any remain alive that long. It will also be a poor lookout for the remnants of the once great herds of Harp Seals which are primarily capelin-eaters and which are still being savagely decimated by Norwegian sealing fleets with some assistance from Canadians.

The prospective elimination of the great base herds of capelin and herring, upon which so many other forms of marine life depend for survival, is actually being used to provide a human rationale for the continuing butchery of seals and whales. This rationale was explained to me by a man who has interests in both whaling stations and sealing ships.

"It is nonsense to try to save the seals, or the whales either. The herring and capelin, and probably squid also, are going to go into the reduction plants so the whales and seals will end up starving anyway. We might as well kill them while they are still some good to us."

Another rationale for continuing the slaughter to extermination was given to me by a fisheries biologist charged with helping to conserve the resources of the oceans: a task which, I suspect, he may feel is already beyond our capabilities.

"A lot of people in the business [resource management] say it's ridiculous to get worked up about saving whales because they are probably going to disappear anyway, because of pollution. All the fish-eating species are at the top of oceanic food chains and they concentrate such pollutants as DDT, mercury and the rest of the stuff we are dumping into the sea. If those things don't kill them outright, it will likely make them sterile, or at least shorten their lives . . . Of course, if we *did* stop hunting them, it would certainly ease the total pressure they are under now and, who knows, they just might make out despite the pollution problem."

There is still another reason for exploiting the remaining whales, of all species and all sizes. This has to do with the extremely heavy capital investment required to build a modern whaling fleet and its ancillary industries. The Japanese, who are now the world's major whalers, claim they cannot afford to stop killing whales until all their investments have been amortized and, *even then, it would be economically wasteful* not to continue to make use of their fleet and plant for whaling since most of it cannot be converted to other uses. Presumably Russia, with the second largest whaling fleet, Norway and other whaling countries such as Canada, agree with this.

However, these are not the publicly stated reasons for continuing the massacre. Propagandists for the whaling interests insist we must continue to kill whales, and in quantity, to provide protein and fat for hungry human beings and to provide industrial and medicinal products vitally needed by modern society. These arguments are, at best, fraudulent. At worst they are downright lies. There is no single product derived from whales which cannot now be synthesized at comparable cost, and proteins and fats can be more effectively produced by farming on a sustained yield basis than by the hunting-to-extinction methods which we still apply to oceanic life.

Nevertheless, we should not be much surprised by any of these attitudes. They are, after all, no more than expressions of the basic approach of modern man toward the world around him. Exploit . . . consume . . . excrete . . . at an ever-accelerating pace. . . . Such is the mad litany of our times.

Like imbecilic children loose in a candy store, we may well come to a sticky end in a belch of indigestion but, if we do not, then we will assuredly die of hunger when the sweets run out. They are running out very rapidly in the oceans of the world. The fond expectation that the seas will feed mankind when the ravaged earth can no longer do so, is no more than an illusion. *Already the seas are being grossly over-fished.* Competent fisheries experts predict, with gloomy certainty, that within two decades food-fish populations in the oceans will have plunged to less than half their present levels, while during the same twenty years the fishing pressures upon them will have increased at least ten-fold!

The appalling destruction of herring in Newfoundland waters is already seriously affecting the inshore fishery for cod and related species. Taken in conjunction with the gross overkill of all the larger species of food fishes on the offshore banks by the burgeoning fleets of trawlers and draggers belonging to a score of desperately competing nations, the loss of the bait fishes (mainly herring and capelin) will mean an early end to *any* significant continuing catch of large edible fishes. It will also mean an end to fisheries as a way of life for many thousands of men. However, not everyone views this as a disaster. As Premier Smallwood, a rather typical politician in the modern mould, once told me:

"That would be a good thing. Yes. A very good thing. A very, very good thing indeed! It would mean the fishermen would have to take jobs ashore as industrial workers. It would lead them into a better way of life, you see. A good thing! Yes, the *best* thing in the world for them."

Chapter 7

Although my attempts to gain insight into the lives of the Finners were bound to be frustratingly inadequate, because I could only encounter them at the interface between air and water, I occasionally had a stroke of luck. Lee Frankham, a friend who was the pilot of a Beaver airplane on floats and who sometimes came visiting and took us joy-riding along the coast, was responsible for one such happy accident.

On a July day in 1964 we flew off with him to visit the abandoned settlement of Cape La Hune, Uncle Art's one-time home. It was a cloudless afternoon and the cold coastal waters were particularly pellucid and transparent. As we were crossing the broad mouth of White Bear Bay, Lee suddenly banked the Beaver and put her into a shallow dive. When he levelled out at less than a hundred feet, we were flying parallel to a family of six Fin Whales.

They were in line abreast and only a few feet below the surface. As seamen would say: they were making a passage under forced draft. Lee estimated they were doing all of twenty knots.

He throttled back almost to stall speed and we slowly circled them. From our unique vantage point, they were as clearly visible as if they had been in air, or we in water, and we could see minute details of their bodies and of their actions. Yet if it had not been that their swift progress underwater was relative to a light

69

wind-popple on the surface, it would have been hard
to believe they were progressing at all.

Their mighty tails and flukes which, unlike the
tails of fishes, work vertically, swept lazily up and
down with what appeared to be a completely effortless
beat. Their great, paddle-like flippers—remnants of the
forelimbs of their terrestrial ancestors—barely moved at
all, for these organs serve mainly as stabilizers and as
diving planes.

There was no visible turbulence in the water although the whales were moving at a rate of knots
which few of man's submarines can equal when submerged. The overall effect was of six exquisitely streamlined bodies hovering in the green sea and seeming to
undulate just perceptibly, as if their bodies were composed of something more subtle and responsive than
ordinary flesh and bones. There was a suggestion of
sinuosity, of absolute fidelity, to some powerful but
unheard aquatic rhythm.

They were supremely beautiful beings.

"It's like watching a fantastic ballet," was Claire's
response. "Perfect control and harmony! They aren't
swimming through the water . . . they're *dancing*
through it!"

Dancing? It seemed a wildly imaginative concept
for I knew these beasts weighed seventy or eighty tons
apiece. And yet I cannot better Claire's description.

Man, being a terrestrial beast of rather rigid per-
ceptivity, is limited in his ability to conceive of alien
beings except in terrestrial terms. In attempting to con-
vey something of the magnitude of the great whales,
men have inevitably compared them with the largest
dinosaurs that ever lived (the great whales are much
larger), or with the largest surviving land animal, the
elephant . . . "a herd of twelve African elephants could
be contained inside the skin of one Blue Whale."

Even more misleading is the concept we have
formed of the great whales from looking at them
stranded on a beach or hauled up on the flensing plan
at a whaling factory. Out of its own element (and stone
dead in the bargain) one of the great whales becomes a

monstrous lump of a thing: a shapeless and gigantic sack, loosely stuffed with meat and guts and fat, only remotely identifiable with the living, functioning entity it once was.

The living whale is something else. The all-too-brief period during which we watched the Fin family crossing White Bear Bay was a revelation. We were all three of us made sharply aware that these creatures were paragons of grace who had achieved a harmonious relationship to the world of waters such as man will never know in air or on the land, in nature or in art.

About ten minutes after we first saw them, the whales rose as one, surfaced, blew and inhaled several times, then sounded while still moving at full speed and without leaving more than a few faint ripples. Since they have excellent vision in air as well as in water, they may have seen our plane. In any event, when they sounded they went deep; shimmering and diminishing in our view as if they were sliding down a long, unseen chute leading to the privacy of the abyssal depths.

Since the last World War men have become very interested in how whales move so swiftly and smoothly in both the horizontal and vertical planes of their three-dimensional world. This interest has not been prompted by admiration or even by true scientific curiosity, but rather by the desire of human warriors to build better submarines with which more effectively to destroy one another. This curiosity, perverted though it be, has led to some fascinating discoveries about the whale as a machine; and everything science has discovered has strengthened the conclusion that whales are among the most highly perfected forms of life ever to dwell upon this planet.

One thing that sadly puzzled early investigators was the question of how a whale could achieve its great speeds with such a "rudimentary" power source as living muscle, and such a simple transmitter of power to water as a pair of flukes. Streamlining was obviously a part of the answer, but only a part. Most modern submarines are almost slavishly streamlined after the whale pattern but, even so, and even when equipped with engines and propellers of maximum mechanical efficiency, they can only achieve speeds comparable to those of whales by expending many times the amount of energy. The secret seems to lie in the fact that the submarine is a rigid object and the whale is not. Experiments in test tanks with those little whales, the dolphins, show that the illusion Claire, Lee and I thought we were experiencing—that of seeing a sort of shimmering undulation in the dancing Fins—was no illusion at all.

Apparently the outer layers of a whale—skin, blubber, and the immediate underlying layers of connective tissue—have the capacity to simulate fluids in motion, almost as if they were themselves a liquid substance. This strange quality produces what hydrodynamic experts refer to as laminar flow, an effect which almost eliminates the normal turbulence produced by an object moving rapidly through water. Laminar flow has the effect (if one can imagine this) of lubricating the whale's body so there is almost no friction or drag. Although I am by no means certain the scientists fully understand what laminar flow is all about, they have certainly recognized its effectiveness. They have discovered, for example, that a dolphin of the same relative mass as a modern torpedo can attain torpedo speed with an expenditure of only one-tenth the energy.

A classic case illustrating the whale's efficiency is referred to by Ivan Sanderson in his fine book, *Follow the Whale*. An 800-horsepower catcher harpooned an 80-foot Blue Whale which then proceeded to tow the catcher, whose engine was running *full speed astern,* a distance of fifty miles at speeds up to eight knots! It was not brute power which made this feat possible. No, it was the almost unimaginable efficiency achieved by the whale's near-total adaptation to the aquatic medium.

We do not know the highest speeds attainable by whales. Some smaller species have been accurately clocked at 27 knots. The misnamed Killer Whale can evidently exceed 30 knots in short spurts, and Fin

Whales have been seen to outrun Killers! However, unlike men, Fin Whales probably do not worship speed as an end in itself. For the most part they seem content just to loaf along, conserving energy, at a modest six to seven knots.

Although we are beginning to learn a little something about the mechanics of the whale as a living machine, we still know very little about the nature of whale society.

The toothed whales, which are more primitive than the baleen, seem to prefer extended family groupings rather like those of baboons and many monkeys. Such groups may include a hundred or more individuals. Polygamy, or at least random mating, seems to be the general rule among the toothed whales. All members of the group or tribe have mutual but generalized responsibilities toward one another. Mature males, or in some cases mature females, may assume leadership roles but all "hands" appear equally concerned with the well-being of the young. When a member of the group is injured, or endangered, all the adults within reach will rally to its assistance. There are many well-authenticated reports of toothed whales physically supporting a sick companion so that it can rise to the surface and breathe. And what is almost unique in the animal world, toothed whales of one species will sometimes come to the assistance of an individual of quite a *different* species.

Among the baleen whales, the social structure seems to be based on the closed family unit. I am convinced that each Fin Whale pod is actually a "nuclear" family consisting of a mated pair of adults, accompanied by the calf of the year plus one or several earlier and as yet unmated offspring. The Fin is not only monogamous; it evidently mates for life, and the bonds between a mated pair are extraordinarily close and tenacious.

Although Finners are strongly family oriented, they are also social in a broader sense. There are reports, dating back to the days when whales were still plentiful, of aggregations of as many as 300 Fins gathered together in one small portion of the sea. These were gatherings of family groups, rather than of individuals. Some whalers believed these gatherings took place two or three times a year and were in the nature of festivals at which unmated whales conducted their courtships in order to establish new family units.

The largest number of Finners seen at Burgeo, after the return of the species to the Newfoundland coast, gathered among the islands during the winter of 1964-65. There were five discrete pods numbering thirty or thirty-one individuals in total. Although some of the families might temporarily come together, and even remain together for a day or two, they would eventually separate again. Each family maintained its own cohesion and each had its own preferred fishing grounds.

The winter of 1964-65 saw the peak and the beginning of the decline of the re-occupation of the south Newfoundland seas by the several species of rorquals. Word of their return had spread all too rapidly until it came to the ears of the Norwegians who, by that time, in company with the British, Japanese, Dutch and Russians, had swept the Antarctic waters almost clean.

Shortly after the end of the war, Karl Karlsen, a financially well-established Norwegian immigrant to Nova Scotia, set up a company to exploit the herds of Harp Seals which drop their pulps—whitecoats, they are called—each spring on the pack ice of the Gulf of St. Lawrence and off the northern coasts of Newfoundland. Karlsen acquired a fleet of sealing ships and built a processing plant at Blandford, near Halifax. In 1964 he expanded the plant to handle whales and, using Norwegian catching ships and Norwegian crews, began going after the rorquals which had reappeared in the seas between Nova Scotia and southern Newfoundland. Soon his big, sea-going catchers, like the *Thorarinn* which is 200 feet long, were ranging five hundred miles and more from the Blandford base, thereby exposing the uselessness of another of the International Whaling Commission's gestures at whale conservation. The Commission had forbidden the use of factory ships in the North Atlantic in order, so it was announced, to provide a sanctuary in mid-ocean for the beleaguered stocks of Atlantic whales. The fact that few of the great whales ever use the mid-ocean reaches but prefer to stay near the coastal shelves where food is much more abundant was not mentioned. In any event, the whole gesture was meaningless in view of the fact that modern catchers, such as the *Thorarinn,* have such great range that those operating from land bases in Norway, Iceland and eastern Canada could, between them, cover almost every area where whales were to be found.

Karlsen's ships worked as far east as the Grand Banks and during their first year took 56 Finners. The next year they killed 108. In 1966 they began to hit their stride, killing 263, and the following year they killed 318. In all, the Karlsen enterprise had killed 1458 Finners up to the end of 1971.* These were in addition to 654 Seis, 64 Sperms and a number of Minkes and Humpbacks. Most of the meat from the more than 2000 great whales so far processed at Blandford has been sold for "animal food," which is to say, for pet food. A good percentage of the oil rendered there has been consumed by the world cosmetics industry, although it is true that some of the meat, and some of the oil, has gone to Japan for human consumption.

With the coming of the Karlsen enterprises, the brief respite during which the great whales had found a sanctuary in the waters south of Newfoundland drew rapidly to a close.

During the winter of 1965-66 only two Fin families returned to Burgeo. There were four individuals in one, and only three in the other, but in addition there was one lone whale whom I believe to have been the sole survivor of a family which had been destroyed by the whalers.

---

*Since 1964, the three stations on the eastern Canadian coast, Williamsport, Dildo and Blandford, have killed 3598 Finners out of a Western Atlantic population estimated at 7000 in 1964. Their total kill of whales of all species in this eight-year period is 5717.

The "Loner," as we called the single whale, spent part of his time in company with one or other of the two family groups, but even more time by himself. Oddly, his favourite fishing place seemed to be the restricted waters of Messers Cove. Here, apparently oblivious to houses, people, and moored boats which almost surrounded him, he spent many hours contentedly eating herring which misguidedly continued to pour into this cul-de-sac. Returning homeward late on winter evenings, I would often hear him blowing in the Cove. He taught me much about his species but perhaps the most surprising discovery was that he had a voice and, moreover, one that could be heard by the human ear, in air.

Late one chilly afternoon I was chatting with Sim Spencer on Messers bridge when we both became aware of a deep thrumming sound which seemed to be as much felt as heard. We turned in surprise toward the Cove and saw a fading pillar of vapour hanging over the icy water.

"Was that the whale?" I asked in astonishment.

Sim's intent face wrinkled in puzzlement.

"Never heard no whale blow like that before. But if 'twarn't he, what do you suppose it were?"

We watched and listened and after a minute or two the sound came again, deep and vibrant; but this time the surface of the Cove remained unbroken. It was four or five minutes later before the whale rose and blew, with no more than his normal whooshing exhalation. Although Sim and I continued to stand there, half frozen, for the better part of an hour, we did not hear that other-worldly sound again. A year was to pass before I would hear it once more and certainly identify it as the voice of the Fin Whale.

While Claire and I were away in the winter of 1966-67, the whales' friends in Burgeo awaited the annual visit of the Finners with foreboding. It was common knowledge that 1966 had been a very good year for the Karlsen catchers and that no great whales had been seen "in passage" by any of our local draggers throughout the autumn months. However, during the first week in December, Uncle Art was delighted to discover they had returned.

It was a sadly diminished band—a single family numbering five individuals.

Throughout most of December these five spent their time, as of old, in the runs among the islands; but during Christmas week their quiet occupancy was challenged by several big British Columbia herring seiners. These great steel vacuum cleaners began sucking up the herring with a relentlessness that was terrifying to behold. Working so close to land that they several times swept away nets belonging to local fishermen, they roused the wrath of Burgeo; but the intruders did not care about the nets, the wrath, or about whales either. On one occasion Uncle Art reported seeing a seiner make what looked like a deliberate ramming run at a surfacing Finner. If it *was* deliberate, then it was also foolhardy, since a collision between ship and whale would have been disastrous to both.

The whales did not take to the newcomers. According to Onie and Uncle Art, they seemed uneasy in their presence. This is understandable since seiners and whale catchers are driven by diesel engines that must sound ominously alike.

A few days after the arrival of the seiners, the Fin family abandoned the island runs and shifted eastward to a little fiord called The Ha Ha, which not even the insatiable seiners dared enter because of many rocky outcrops that might have damaged their costly nets. The whales stayed close to The Ha Ha and the nearby mouth of Bay De Loup, except when the seiners were absent delivering their catches to the reduction plant at Harbour Breton. It was on one such occasion, when only a single seiner was present, that Uncle Art and I stood on Messers Head and watched the whales demonstrate the superiority of their fishing techniques.

The whales were not alone in The Ha Ha. They shared it with several Burgeo fishermen working cod nets from open boats. When the whales moved in, these men were concerned for the safety of their nets.

Two among them, the Hann brothers, Douglas and Kenneth—small, quiet, foxy-faced men from Muddy Hole—even considered moving their gear to some safer ground.

" 'Twarn't as we t'ought they'd tear up our gear a-purpose-like," Douglas Hann remembered, "but The Ha Ha is a right small place and not much water at the head of she. We t'ought, what with six fleets of nets scattered round, them whales was bound to run foul of some of them . . . couldn't help theirselves. Well, sorr, they never did. Sometimes when we'd be hauling a net they'd pass right under the boat close enough you could have scratched their backs with a gaff. First off, when they did that, we used to bang the oars on the side of the boat and yell to make them veer away; but after a time we sees they knowed what they was about, and was going to keep clear without no help from we.

"Still and all, 'twas scary enough betimes. One evening our engine give out. We had the big trap skiff and no thole-pins for the oars so we had to scull her along. It were coming on duckish* and we was alone in The Ha Ha and them whales begun coming up all round. They was only six fathom of water where we was to, and they was after the herring like big black bullets. We could hear the swoosh when they drove by, and foam would fair bile up where they took a big mouthful out of a herring school.

"I'd as soon have been home in me kitchen, I can tell you, but them whales is some smart navigators, for they never come nigh enough to do we any hurt. We was an hour poking our way to the pushthrough what leads into Aldridges Pond, and them whales stayed right along of we. Toward the end of it they give up fishing and just come along like they knowed we was into some kind of a kettle. Ken, he said maybe they was offering we a tow; but I suppose that's only foolishness."

What the Hanns told me of their experience reminded me of a story I had heard some years earlier from a very old man at Hermitage Bay, many miles to

*Meaning dark.

the eastward. As a youth this man had been employed at a whale factory in Gaultois, on the north shore of Hermitage Bay. His home was five miles across the full breadth of the Bay, but on weekends he would row over to spend Sunday with his family.

One Saturday afternoon he was homeward bound when he saw a pod of Finners. There were three of them, and they were behaving in a peculiar fashion. Instead of briefly surfacing and then sounding again, they were cruising on the top. Their course converged with his, and as they drew close, my friend saw that they were swimming, as he put it, "shoulder to shoulder." The centre whale was blowing much more rapidly than the rest and its spray was pink in colour.

" 'Twarn't hard to know what was the trouble," the old man remembered. "Yon middle whale had been harpooned and the iron had drawed and he'd got clear of the catcher boat. The bomb must have fired, but not deep enough for to kill he.

"I laid back on me oars, not wantin' to get too handy to them three, but they never minded I . . . just steamed slow as you please right past me boat, heading down the bay and out to sea. They was close enough so I could near swear the two outside whales was holding up the middle one. I t'inks they done it with their flippers. That's what I t'inks.

"When they was all clear, I rowed on home. Never t'ought no more about it. About the middle of the week a schooner puts in to Gaultois and the skipper was telling how he come across three whales outside Green Island. They was right on top of the water, he says, and never sounding at all, and making a slow passage to eastward. The skipper, he held over toward they, but when it looked like they was all going to go afoul of one another he had to alter course, for they had no mind to sound no matter what he did. When he passed alongside he saw the middle whale had a girt hole in his back.

"I made certain 'twas the same three I come across and 'twas agreed by all hands 'twas the same whale was harpooned by one of our boats that Sunday morning and got away. They two other whales took the sick one off someplace . . . some said 'twas to the whales' burying ground . . . but all I knows is they kept that sick one afloat somehow for five days, and close onto sixty miles."

Aldridges Pond is a salt-water enclosure about half a mile in length, and nearly as broad, lying in the centre of the rocky isthmus which separates The Ha Ha from Short Reach. There is a narrow and very shallow "pushthrough" between it and The Ha Ha, passable only by small boats and then only at high water. However, a wider and deeper channel connects the Pond to Short Reach by way of a rather large entry cove. It was the habit of the men who fished The Ha Ha to pass back and forth through Aldridges Pond to save themselves the long and, in dirty weather, dangerous outside run around the head of the peninsula. Each morning at daybreak they would cross Short Reach, enter and cross the Pond, pole through the pushthrough, and set to work hauling their nets in The Ha Ha. In mid-afternoon, when the haul was finished, they would bring their loaded boats back into Aldridges and moor up to the shore in the Pond's protected waters to gut their catch.

During our years in Burgeo, Claire and I had only once visited Aldridges Pond; but before we had been home two weeks, the Pond became the centre and the setting for an event which was to change our lives.

Chapter 8

The weatherman was predicting a sou'easter for Friday, January 20—the worst kind of storm on the Burgeo coast—and the Hann brothers were in a hurry to get clear of The Ha Ha. Working under a lowering sky with snow squalls flickering around them, they hauled their last net and cleared it shortly after noon. Beating the ice from their homespun mittens, they cranked up their little 5 h.p. Atlantic and began chugging homeward toward the pushthrough leading into Aldridges Pond. They had seen no whales all day and, as they cut the engine and began poling the boat through the shallow entrance, Kenneth remarked to his brother that the whales must have heard the weather report too.

"They knows the seiners has sheltered up, I wouldn't doubt. It'll be clear fishing for them whales 'mongst the islands."

"They's welcome to it," Douglas replied, with a concerned glance at the ominous overcast.

The Hanns let their boat drift westward across the Pond while they hurriedly gutted the small lot of fish they had caught. Within an hour they were finished. As their boat puttered out through the south channel into the entrance cove which leads to Short Reach, they found themselves bucking a heavy tidal stream that was racing back into the Pond. The old boat swayed and yawed uncertainly back and forth across the channel until Kenneth had to fend her head off with an oar.

As he was doing so, he noticed that the water around him was alive with herring.

"Don't say as I ever see them so thick," he recalled, "pretty near enough to float our boat on their backs. They was pouring into the Pond like the devil was on their tails. Maybe he was, cause just after we got into Short Reach we see two spouts just abeam of Fish Rock. It looked to we like the whales was having themselves some right good fishing."

During the previous several days the Hanns had noticed that, despite its almost landlocked nature, Aldridges Pond was becoming increasingly attractive to the herring. They laid this to the fact that, hard-pressed by the seiners, the little fishes were being driven to seek refuge where the ships could not follow. Whatever the reason, the Pond had acquired a dense herring population which seemed to surge in and out with the ebb and flow of the tides.

Despite the ominous forecast, it only blew a "moderate breeze" that night. The anemometer on our roof registered forty knots and when Claire and I went to bed we could feel the full reverberation of breakers exploding against the cliffs of nearby Mast Cove, but it was a brief blow. Before morning the wind had hauled into the north, the skies were clearing, and the sea was falling out.

The Hann brothers, hard-working men with big families to support, were early getting out to haul their gear and they entered Aldridges before dawn began to break. They noticed nothing unusual as they thumped across the Pond, but as they were leaving the push-through they encountered several whales clustered near that narrow entrance. Vaguely Kenneth wondered if they might be waiting for herring to come out with the falling tide, but he was too concerned just then about his gear to give the matter much thought.

The brothers were delighted to find their nets undamaged and even more delighted to find them full of cod. They spent several hours clearing and resetting the nets and when, heavily laden, they started back for the pushthrough, they again passed close to several whales surfacing near the entrance. Paying them no particular heed, they poled through the channel, moored their boat to a rock on the north shore of the Pond and began gutting their catch.

They had been at work only a few minutes when they were startled by what would have been a familiar enough sound outside but which was totally unexpected in the confines of the Pond. It was the explosive *whoooof* of a spouting Finner; and it came from "close aboard." When Kenneth, a voluble and excitable little man, described the discovery some time later, he was still amazed.

"I tells you, we was some surprised. We looked up and not more'n a couple of chains off was the fin of a girt whale just slipping down into the water, and the fog from her spout still hanging in the air.*

" 'Twas hard to believe a whale that big could get herself into the Pond. Certainly there was no water for her in the pushthrough, and even the south gut never had no more'n two fathom and a half since ever I see it. But there she was, and looking twice as big as life.

"We went on cutting our fish, but I tell you we kept a sharp lookout for that whale. She was down for quite a time and when she come up again she was going like a battleship and right on course for the south gut. I give a yell: 'Doug, bye! She be going to drive herself ashore!'

"We both stood up to watch her. She never slowed 'til she was fair in the mouth of the gut, and I made sure she was going to end up high and dry. But just at the last of it she changed her mind. She come about so hard she sent a wall of water right out the gut. If they had been a boat inbound, 'twould have been capsized for sure. She turned so quick she heeled right over on her side, and she was *some* big! We figured sixty, seventy feet and maybe more."

---

*The Hanns did not realize that the whale actually *was* a female, but so impressed were they by her gigantic size that they automatically referred to her in the feminine gender exactly as they would have done had she been a ship.

During the next hour, while the Hanns finished cutting their fish, the whale went on making these frantic and abortive rushes, as if trying to nerve herself to dare the shallow channel. Time and again she surfaced near the middle of the Pond, put on a burst of speed and headed for the opening. But each time she gave up the attempt at the last moment. She must have known that not even her great speed and momentum would suffice to carry her across the shoals which barred the mouth of the channel.

"I wondered," Doug Hann recalled, "was she sounding the rising water 'cause the tide was on the flow. 'Twas near sprinjes* and we thought maybe she might get clear on the top of high water that evening."

Having finished with their fish, the Hanns found themselves in something of a quandary. They had no desire to be caught in the south channel during one of the whale's impetuous rushes. On the other hand, they did not much relish the idea of taking their heavily laden boat back into The Ha Ha and out around Aldridges Head in the teeth of the heavy swell which was the storm's aftermath. Eventually they decided to ease their way toward the south channel, holding close to shore, and see what happened.

The whale paid no attention to them until they were within a hundred yards of the channel mouth. Here they discreetly nosed their boat ashore, uncertain what to do next. Then the whale surfaced in mid-Pond and turned in their direction.

"What happened next was the queerest thing I ever see," Kenneth remembered. "Whenever I see a whale spout before, and I supposes I see a t'ousand, only the top of the head and a bit of the fin and back would come out of the water. But this one . . . she made a slow start toward me, went under, and next thing we knowed was standing up with her whole head clear of the water, high as a cliff, and staring straight at we out of one big eye.

"I tell you, it scared me some! Her mouth was

---

*Spring tides, the highest tides of the lunar month.

closed but we could see 'twas big enough to swallow a dory and room to spare. Then she slid back under. We waited and waited but she were gone pretty near twenty minutes and when she come up again she were at the north end where we'd been cutting fish. I says to Doug: 'Start the engine, bye! I t'inks she's going to give we a free passage out.'

"We never wasted no time, I'll tell you. We was out through the gut like a rat out of a red-hot stovepipe! Going back down The Reach we talked about how wonderful queer it was for that whale to get herself into the Pond. 'Twas Doug had the best idea of it: ' 'Twas the herring took she into it,' Doug says. ' 'Twas the herring tells the tale!' "

Although the cove from which the south channel leads into Aldridges Pond is small, it is surprisingly deep, with an average depth of thirty feet over most of its area even at dead low tide. However, at its head, toward the Pond end of the channel, it shoals abruptly to a low-tide depth of five feet, and it narrows equally abruptly to form the mouth of the channel which is itself some forty yards in length and ten yards wide. Near the inner end of the channel are a number of huge but scattered boulders, some of which are covered by less than four feet of water at low tide.

On Friday night, January 20, the predicted tidal rise was just under five feet but the actual rise was nearer six due to the heavy sea driving inshore before the brief sou'east gale. Consequently, at full of the tide that night, there was close to forty feet of water in the cove and ten or eleven through most of the channel itself.

Early Friday afternoon the half-dozen seiners which had been working among the Burgeo Islands began to stow their gear and head for port to ride out the coming storm. The last of them left at about 3 o'clock and the underwater world no longer vibrated to the pound of diesel engines.

With the return of quiet to their world, the family

of Finners in The Ha Ha decided to try a change of ground. Perhaps the herring had begun to avoid the restricted waters of The Ha Ha which were being so closely patrolled by the whales, or perhaps the five great, hungry maws had simply thinned them out. In any event, the whales rounded Aldridges Head and entered Short Reach where two of them were seen by the Hann brothers as they headed up for home that evening.

Short Reach, it happened, was alive with herring, and the whales had good fishing while daylight lasted. However, dusk came early this Friday evening as the cloud came down, dark with the promise of the coming storm.

With the coming of night, the preferred method of fishing, using reflected light from the abdomen to herd the herring, became ineffective, but by then most of the whales had gorged and were no longer hungry. The *pater familias,* a sleek giant only slightly smaller than his mate, and three young but well-grown progeny of other years, were content to idle in the depths and wait for another day. Not so the female. She was still ravenous, and she had good reason for her gargantuan appetite; for she was carrying a calf within her vast womb, a new life growing at such a furious pace as to keep her perpetually hungry. So while the rest of her family took life easy, the female continued fishing in the dark, relying now on the high speed "dash," directed at the densest herring schools she could find.

Locating such schools posed no great problem because her sonar told her where they were, how dense the schools were and the depth at which they swam. Since she could not use the herding technique, it was to her advantage to choose those schools which would be prevented, by some natural barrier, from scattering at her approach. The many coves and cul-de-sacs along the steep, deep shores of the Richards Head peninsula were well suited to her purpose and it was not long before she discovered that one of them, the entrance cove to Aldridges Pond, was fairly seething with little fishes.

Her sonar gave her sight in darkness but it was not perfect vision. Although quite capable of defining for her the shape and depth of the cove, it probably did not reveal the existence of the narrow, shallow channel leading into Aldridges Pond. Unaware, then, that the packed mass of herring had an escape route open to them, the female whale came streaking into the cove like the black shadow of nemesis . . . only to find that, inexplicably, the school was fading away before her. She would have had no way of knowing that it was pouring like a living river through the channel to seek safety in the pond.

The cove was deep but it was dangerously small. If the whale had accepted the fact that her attack had somehow failed, she could have halted her rush in time by gaping her enormous mouth and allowing the inrush to distend her accordion pleats so that the resistance of the water would have brought her to a halt nearly as abruptly as if she had deployed a parachute brake. But the hungry female fractionally delayed applying her brakes. When she did decide to stop, momentum carried her into the mouth of the channel where the water abruptly and horrifyingly shoaled until her belly touched the bottom rocks. She swung frantically back and forth trying to turn, for a whale cannot swim in reverse. But assisted by the fast-flowing inbound tidal current, her struggles only served to carry her farther into the channel.

Now she must have known fear; for a great whale that goes aground at high tide is almost certainly doomed. As the tide ebbs, the tremendous body, no longer buoyed up, becomes its own instrument of death. The ribs collapse under the immense weight and the whale suffocates.

Caught in the tidal stream, and propelled by the frenzied thrusting of her flukes, she moved in the only direction open to her—forward through the channel, contorting herself to squeeze over and between the boulders which partially barred her way. Then, miraculously it must have seemed, the water deepened and she was free.

There is an average depth of five fathoms over most of Aldridges Pond, and nine in the middle. The whale's relief on reaching apparent freedom must have been exquisite. It must also have been short-lived. It could not have taken her many minutes to scan the shoreline and to realize that she had escaped . . . into a trap; a trap from which there was no hope of release except through the channel by which she had entered.

Then came a fearful moment of decision! To stay was, eventually, to starve; but to dare that channel on a falling tide almost certainly meant death, a quicker one, before the next day came.

Of course the tide would rise again. Certainly she would know that. But that was not all—and not enough—for this was happening at spring tide and the water would not be so high again for the whole of a lunar month and, even then, it might not be high enough unless accompanied by an onshore gale. Perhaps she knew that too.

The inexorable rhythm of the tide was the whale's key to freedom. Throughout the rest of that Friday night she waited for the new tide to rise. It reached its peak late Saturday morning . . . and the peak was almost a foot below the level it had reached the night before. It was not enough. Nevertheless, she was inexorably drawn to the mouth of the narrow channel which was her only path to freedom.

The Hann brothers reached the plant about 4:30 in the afternoon and began forking their fish up on the wharf, while an interested audience of plant workers listened to the tale they had to tell. It provided a major diversion in the otherwise dismal routine of working in a fish packing plant.

"You think she's still into the Pond?" asked one of the foremen, a strapping-big—almost obese—man in his fifties.

"More'n likely. Don't see as she got a chance get clear until full tide tonight."

"But," Kenneth Hann told me later, "if I'd a

knowed what they fellers had in mind, I'd have told they she was gone clear of it already."

Perhaps he would have done so, for the brothers wished the whale no ill. There were other men in Burgeo who did not feel as the Hanns felt.

Chapter 9

Only a few minutes after the Hanns finished unloading their fish and departed for their homes in Muddy Hole, another boat put out from the plant wharf carrying the foreman, George, who had been so interested in the whale, together with four younger men. All five were employees of the plant and they had obtained permission from the manager to quit early in order to go and "see" the whale. They did not go straight to Aldridges Pond. Partway down Short Reach they put in to shore and each man hurried up to his own house, returning with a rifle and whatever ammunition he could lay hands on. When they shoved off again, they were armed with two .30.30 sporting carbines, a .30.06 U.S. Army rifle, and a pair of .303 Lee Enfield service rifles.

Although these men had all been born on the Sou'west Coast, they had all spent some years away, either in Canada or the United States. Returning home for one reason or another, they had rejected the vocations of their fishermen forefathers and had instead sought wage employment at the plant as mechanics, tradesmen and supervisors. They were representative of the new Newfoundlanders envisaged by Premier Smallwood—progressive, modern men who were only too anxious to deny their outport heritage in favour of adopting the manners and mores of 20th-century industrial society.

It was already growing dusk when their boat passed Richards Head and opened the mouth of the cove leading to Aldridges Pond. Suddenly the man in the bow gave a shout, grabbed his rifle and pointed with it. A whale was spouting just off the entrance of the cove.

The foreman throttled back and the boat slowed to a drift.

"Son of a bitch must'a got clear!" someone cried disgustedly.

As the whale sounded, three or four high-velocity bullets whacked into the water and ricocheted, spinning and singing into the darkening air. Although the boat's crew waited expectantly for the great beast to reappear, it did not do so. Believing it to be the whale reported by the Hanns, and convinced it was now well on its way to sea, the disappointed men decided to head home. George, the foreman, was turning the boat when one man, who had only recently returned to Burgeo after a decade spent working in a Toronto paint factory, spoke up:

"I never been into the Pond since I was a kid. Let's take a turn. Might stir up a couple of sea ducks or maybe a seal."

Agreeably, George completed a full circle and the boat headed into the cove, crossed it, and nosed into the channel. At that instant all five men were electrified by what they later described as a sort of hollow roar—"like a cow bawling into a big tin barrel." Almost at the same moment, there was a mighty splash at the inner end of the channel and a drift of spray descended on the boat and its occupants. The men stared at each other in amazement.

"Holy Jesus Christ! 'Tis still in there!"

They ran the boat's nose ashore on a patch of shingle, sprang over the side and raced across the narrow neck of land separating the cove from the Pond. They reached the crest in time to see what one of them described as ". . . the biggest goddamn fish that ever swum! It was right into the mouth of the gut and looked to be half out of the water. The tail onto it was like an airplane. It was smashing into the water so hard, spray come right up where we was to."

The five men wasted no time. Some dropped to their knees, levering shells into their rifles as they did so. Others stood where they were and hurriedly took aim. The crash of rifle fire began to echo from the cliffs enclosing the Pond and, as an undertone, there came the flat, satisfying thunk of bullets striking home in living flesh.

"You couldn't miss," one of the gunners recalled ecstatically, "I was shooting for the eye—big as a plate it looked. Some was shooting for the blowhole. But them bullets only seemed to tickle it. It kind of rolled a little, turned and slid off into deep water and went down. I guess we put twenty bullets into it."

The whale submerged and retreated to the north end of the Pond but after a few minutes, during which the gunners hurriedly reloaded, she reappeared heading back toward the channel mouth. The men held their fire until she was a hundred feet away and then let fly a concentrated blast. The whale sounded again, this time disappearing for nearly twenty minutes.

"Some of the lads figured as how we'd killed it, but that was crazy stuff. Take a hell of a lot more'n what we had to kill a thing that big! But George, he figured if we kept shooting every time it come toward the gut it'd give up on the idea and stay where it was to until next day. Next day being Sunday we'd have all day for sporting with it."

The men had not brought much ammunition and after an hour had exhausted their supply. Reluctantly they returned to their boat and to Burgeo where they spent the rest of the evening in convivial drinking as they toured from house to house around The Harbour and The Reach, telling the story of their exploit.

Sunday broke white and clear, with blinding sunlight glaring from the snow and ice of a silent land and on the sheen of a still sea. Completely unaware of what had been happening at Aldridges, Claire and I and Albert set out to the westward for a day-long hike on the scimitar of sand beaches which bars the waters

of Little Barasway Pond from the open ocean. It was
an unforgettable day for we saw seventeen Bald Eagles
—an almost unbelievable number in this age when the
Bald Eagle has become yet another vanishing species.
They had gathered on the beaches from God alone
knows how many hundreds of miles away to feed on
tens of thousands of dead herring—victims of the
seiners—which had washed ashore.

Meanwhile in Burgeo the bells of the Anglican
and United Churches were ringing and men and wom-
en were making their ways along the icy roads to
morning service. After the services, knots of men gathered here and there to yarn a bit before going home for
Sunday dinner. The main topic was the trapped whale.
It would have been difficult to ignore the subject because ever since early morning a distant obbligato of
rifle fire had been audible east of the town.

Not long after dawn, between twenty and twenty-
five gunners had ringed Aldridges Pond. This time
each man had a good supply of shells. A number of
the merchants had obligingly opened up their stores
and soon sold most of the high-velocity ammunition
they had in stock. The morning shooting party included
several of these merchants, together with some staff
members from the plant and a good representation
from among the smart young fellows who spent their
springs, summers and autumns working on the Great
Lakes freighters, and their winters at home in Burgeo
drawing unemployment insurance. Also represented
was Burgeo's first businessmen's organization, the recently formed Sou'westers Club, which was a combination service group and Chamber of Commerce committed to the proposition that Burgeo must transform
itself as rapidly as possible into a really modern town.

With the arrival of the sportsmen, the whale retreated to the middle of the Pond where she spent as
much time as she could submerged.

The system whales have evolved to enable them, as
air-breathing mammals, to survive for long periods in
the depths is wonderfully effective. Unlike a human

diver, the whale does not rely solely on the air it can store in its lungs. If it were to dive with lungs fully inflated, it too would be subject to the crippling and often fatal effects of what we know as the bends. Instead it stores most of the oxygen it needs in the red corpuscles of the blood and, by means of a special chemical process, in the muscle tissues too. Furthermore, a whale on a deep or prolonged dive restricts the blood circulation so that the precious oxygen is only distributed to those organs and those regions where it is most vitally needed.

When a Fin Whale surfaces after a long dive (there are records of dives of at least 40-minute duration), it must expel the waste gases which have accumulated and then take in enormous quantities of fresh air; and it must do this swiftly, for it is vulnerable when it is at the surface of the sea. The arrangements which make this possible are impressive.

A whale's blowholes are actually a mammal's nostrils, which have migrated from the front to the top and back of the head. They are equipped with powerful valves and are connected to the lungs by a huge passage capable of handling an immense volume of air. The great bellows of the lungs are so effective that when a Finner surfaces it can totally exhaust and then refill them to the bursting point in just over one second! After a long submersion this breathing rhythm must be repeated several times before the depleted supply of oxygen is fully replenished, and there must be an interval between breaths to allow the blood to absorb the oxygen which has been taken into the lungs. A Fin Whale submerges after each separate breath, rising to spout again at intervals of two or three minutes until, fully "recharged" with oxygen, it can again return to the deeps.

The first part of a surfacing Finner to become visible is usually the hump housing the blowholes. The moment the hump breaks surface the valves snap open; the whale lets go an explosive exhalation which is instantly followed by an implosive indraft. Then the valves snap shut and the great beast begins to descend, appearing to cartwheel forward in a flat arc so that the blowholes are followed at the surface by a long expanse of curving back and finally by the high fin, which is set well aft. The flukes seldom show, but twin circular disturbances of roiling water testify to their powerful presence as the whale vanishes from view.

The brief surfacings of the whale did not give the sportsmen at Aldridges Pond much chance to spot their quarry, take aim and fire, and at first the gunners did not make very good practise. However, they soon began to recognize a pattern in the whale's appearances and to take advantage of it. When she first surfaced after a long submergence, they held their fire. When she rose again to take her second and subsequent breaths, they were ready for her. The rifles cracked and the echoes reverberated back and forth from nearby Richards Head to Greenhill Peak.

As the day wore on, more and more boats arrived and an atmosphere of fiesta began to develop. Most people were content just to watch the show from the natural rock amphitheatre which cradles Aldridges. Men, women and children, many in their Sunday best, sat or stood and watched with eyes that brightened when the whale, beginning to panic, turned into shoal water, touched bottom, reacted in terror, and then flailed her way back to deep water again while a continuous *BAM-whap . . . BAM-whap . . . BAM-whap* told how well the sportsmen were doing.

Not everyone who saw the show that day was happy about it. An elderly fisherman from Muddy Hole, who had brought his daughter and grandchildren to see a live whale, was disconcerted.

"It looked like rare foolishness to I," he said. "What war the use of it? Them bullets cost money and they was heaving 'em away like they was winkle shells. 'Twould have been better to save their bullets to get a bit of meat in the country . . . but I supposes the likes of them chaps got more money 'an they needs.

Still and all, they had no reason to torment the whale. She war no good to they. Happen they killed her, what was they going to make of her? Put her through the plant for frozen fillets? No, me son, 'twas a pack of foolishness."

Another friend of mine, the mate of a dragger, was there that day. He did no shooting either and his contempt for those who did was surprisingly explicit in a community where criticism was seldom openly expressed.

"There was three young chaps there I knowed for what they was . . . bloody butchers. March of last year when the glitter drove the deer out to Connaigre Bay, they come along to have some fun. Nothing better to do, I suppose. I was aboard the *Pennyluck*, anchored up the Bay with engine trouble. When them chaps come, Lard Jasus, I thought 'twas the Germans. Shoot? They never stopped, sunup to sundown. I was ashore next day and they was dead deer, cows and bulls alike, all over the place. Yiss, bye, I knowed them chaps. What was their names? That's for me to know and you to find out. Rotten bastards they is, but I'll not tell their names."

I changed the subject and asked him to describe the whale's behaviour under fire.

"It made me feel right ugly just to watch her. She had no place to go, only down under, and she couldn't stay down forever. I expected her to go right crazy after a couple of hundred bullets had smacked into her, but it was like she knowed that would do her no good. Once or twice she got wild, but for the most of it she war right quiet.

"One thing . . . she warn't alone. I was standing high up on a pick of rock where I could spy into the Pond and out across Short Reach as well. 'Twasn't long afore I sees another whale outside. It blowed first just off Fish Island. Then there was a stretch of time when no boats was in The Reach, and it moved closer and closer until 'twas fair in the mouth of the cove.

"Now here's the queer thing. Every time the whale

in the Pond come up to blow, the one outside blowed too. It happened *every* time they blowed. I could see both of them, but they was no way they could see one t'other. You can say what you likes, but the one outside knowed t'other was in trouble, or I'm a Dutchman's wife.

"They fellows shooting at the whale weren't no smarter nor a tickleass*. They was using soft-nose bullets and when they hits something they busts up . . . goes all to pieces. That whale must a had a foot of blubber and I don't believe the most of them bullets got through at all. I supposes some did, certainly, but I don't say she was hurted bad. Try and kill a thing that big with soft-nosed slugs? My son, they might as well have took to heaving rocks at her!"

The shooting gallery went out of business early on Sunday afternoon when the supply of ammunition ran out. With nothing more to entertain them, most of the onlookers went home. My friend was one of the last to leave.

"The last I see of the whale, she was swimming pretty good. Was no blood coming from her blowhole but she weren't blowing near so high nor staying down so long. Maybe she was just played out." He paused thoughtfully and then concluded in a tone that indicated some embarrassment. "A man didn't feel right about what was done that day. 'Twas no great credit to us folks in Burgeo. . . ."

When Claire and I returned from our walk late Sunday afternoon, we were tired and content. The tensions which had beset us during the long months of travel had sloughed off like old skin. The outer world of turmoil and disaster now seemed so distant as to be almost irrelevant. I do not ever recall having felt more at ease within myself than I did that night as we prepared to go to bed.

Claire wrote in her journal:

---

*The local name for the small gull otherwise called kittiwake.

"It is so nice to be back among people who live simple and uncomplicated lives. We have really missed the people here. How I hope they are never spoiled by the savagery and selfishness that seems to be spreading over the whole world like a fog. . . ."

Chapter 10

By Monday morning a bitter nor'easter, blowing down from the frozen barrens, had shrouded Burgeo under a low and scudding overcast. Those who could stay at home by their kitchen stoves did so gladly. Few fishermen were out; but the plant workers, men, women, and children as young as fourteen, red-faced and bending to the chill wind, shuffled off as usual to begin their bleak day's work. They stood for endless hours on cold concrete while numbed hands filleted and packaged a thin stream of fish coming along a conveyer belt from the belly of a deep-sea dragger unloading at the wharf.

There seems to have been little talk about the whale that day. Those who had been at Aldridges on Sunday were disinclined to discuss the matter. A letter written to me by one of the Burgeo clergymen a year afterwards may throw some light upon this reticence.

"I hope you won't think all the people were acting in their ordinary way that Sunday. Afterwards a lot of them felt very bad about it. Except for those few who did the shooting, most of the people had no idea of harming the whale. When they saw the shooting and watched it for a while they got worked up by it; and afterwards a lot of them felt very bad. They hardly knew how to tell what was the matter, but most of them was quite upset next day. . . ."

That there was indeed something like a conspiracy of silence explains what is otherwise difficult to under-

stand; that Claire and I, living less than three miles
from Aldridges Pond, heard not a whisper about what
had taken place there. Onie Stickland, Simeon Spencer,
Uncle Art, and the rest of our friends and neighbours
who were usually so ready, not to say eager, to keep us
informed of everything that happened in Burgeo, spoke
not a word to us about the whale.

On that Monday morning, while we remained in
ignorance, the Hann brothers were cautiously and
somewhat fearfully entering Aldridges Pond. As Kenneth recalled it:

"We knowed the gunning *might* have killed the
whale, though 'twas more likely she was just bad
hurted. But you got to watch out for even a deer with
a bullet in its guts, and that whale, she had hunnerts of
bullets into she. There was no telling what she *might*
do, but we knowed one flip of her tail could put an
end to we."

Gingerly they puttered through the south channel,
stopping just inside to see what the situation was. The
surface of the water was ruffled only by the wind. Although they waited a quarter of an hour they saw no
sign of the whale. So they started up their engine and
headed across the Pond, but they had gone only a hundred yards when she surfaced close ahead and almost
on a collision course. She blew once and began to
sound. The two men frantically snatched up their oars
to help the engine push them to the safety of the shore.
They were rowing for dear life when they were appalled
to see the enormous head passing directly beneath their
boat.

"I says to myself, 'Kenneth, me son, that's it!
You're gone out now!' I was certain sure that whale,
with all them bullets into her, would be so hateful she'd
lay into the first humans she could reach. Feared? I
was shaking like a dog!"

But Kenneth was wrong. The whale passed ma-
jestically on her way and when she surfaced again it
was at the far end of the Pond. Still shaken, the Hanns
hastened to the pushthrough and on into The Ha Ha
where they spent the next several rough and frigid

hours wrestling with their gear. They were glad to regain the shelter of the Pond but they were still leery of the whale, so they prudently pulled their boat to shore close to the entrance before beginning to gut their fish.

While they were busy at their task, the whale continued to cruise steadily around the deep part of the Pond, blowing at intervals of about ten minutes and showing no further desire to dare the passage of the south channel. Once or twice the men saw big, circular water boils appear on the surface and they thought the whale was chasing herring.

Such placid behaviour restored the Hanns' confidence somewhat and as they coasted the shore of the Pond, homeward bound (and keeping, as they say, "one foot on the land"), Douglas was moved to take a bait-tub full of herring and empty it into the water not far from where the whale had just submerged.

"Don't know as she cared for it," he explained, "but I thought 'twould do no harm. Truth to tell we was feeling friendly toward she . . . she give us a clear passage through the Pond, spite of what them fellers done."

As far as is known, the Hanns were the only people to visit the whale on Monday . . . but Tuesday was a different day.

The weather had improved by Tuesday noon and a party of riflemen, hearing that the whale was still alive, thought there might still be some sport to be had with her. The only difficulty was a shortage of ammunition, but there was an answer to this problem. Like many remote Canadian communities, Burgeo boasts a platoon of Rangers, a semi-military organization of volunteers under command of one of their own number who holds a temporary commission in the Canadian armed forces. Each Ranger is issued with a .303 service rifle; and cases of ammunition are kept at each detachment headquarters. Part of this ammunition is issued for target practise while the balance is retained for use in case of a "military emergency." The Burgeo Rangers had long since exhausted their practise issues, mostly on caribou, moose and harbour seals.

It happened that several of the whale hunters belonged to the Ranger platoon. On Tuesday morning one of them visited the second-in-command of the detachment, who was also a senior member of the fish plant staff, and asked for a special ammunition issue. The spokesman did not claim there was an emergency in Burgeo but he did point out that target practise never came amiss and that it was unlikely they would ever find a better target than "that whale, up to Aldridges."

A number of rounds were issued . . . just how many I was never able to determine, but I was later to count more than 400 empty .303 cartridge cases lying in piles around Aldridges Pond, all bearing the markings of Canadian army arsenals.

Apparently Tuesday's gunners were not entirely easy in their minds. Perhaps they were aware that an undercurrent of disapproval now existed in the community. Or possibly they were a bit worried about the illegality of what they were doing since, of course, they knew that even the carrying of rifles, let alone their use, in the countryside at this season of the year was prohibited, as a measure intended to protect the caribou and moose from poachers.

For whatever reason, those who manned the three boats heading down The Reach toward Aldridges late Tuesday afternoon had waited until all the fishermen who used the Pond had left it. In fact they were so circumspect that nobody seems to have even seen their boats depart. Had it not been for the unexpected arrival on the scene of a man and his son who, in their dory, had been fetching a barrel of spring water from a stream at the foot of The Ha Ha, the names and activities of Tuesday's sportsmen might never have been known. Of the eleven gunners who disembarked at the Pond that day, eight had taken part in Sunday's affray, while three were respected members of Burgeo "society" who had missed the Sunday fun because of church commitments.

The man who had gone to fetch water at The Ha Ha was not hesitant about describing what he saw.

"We heard the shooting before ever we left the spring, but never thought much about it till we come to the pushthrough. The young lad was up in the bow and we no sooner got into the Pond when a bullet comes wheening over his head, close enough so he could feel the wind.

"Well, sorr, we hauled the dory in behind a rock, right smart. Then I takes a spy over the top. 'Twas something like I never see before. There was that crowd from down The Reach, one of them fellers on every point, and carrying on like they had all gone right foolish. They was yelling and jumping, and hauling away on bottles, and shooting at the whale, and yelling some more. There was bullets flying every which way. 'Twas a wonder some of them never killed the others."

What he was seeing that Tuesday afternoon was essentially a repetition of Sunday's fusillade . . . with one important difference for the whale. The army-issue bullets used on Tuesday were *steel-jacketed* and so had a penetration far greater than that of commercial, soft-nosed bullets. Instead of shattering into small fragments after making a relatively shallow entry into the whale's blubber, these bullets pierced deep into her body.

"The creature was drove clean crazy! She got herself into the shoal water on the eastward side of the Pond, the farthest she could get from them fellers, and there warn't no more than just enough water to keep her afloat. And beat the water! My dear man, her tail and flippers was flying in the air! 'Twas a desperate sight, I tell you. And all the while you could hear the thump of them bullets just a-pounding into her.

"After a time she kind of thrashed into deeper water and then she took a run for where we was to. I swears to God I thought she was going to come right on top of we! She come to a stop fair in the mouth of the pushthrough with her mouth swole up like one of them balloons they used to fly on ships in the war to keep the German planes away. I tell you, 'twas more'n I cared to see! The young lad and me, we hauled back into The Ha Ha and rowed right out around Aldridges Head. I'd a rather crossed the ocean in me dory than gone across the Pond that evening!"

This man had one further detail to add. As he and his son were rowing into Short Reach they encountered four more whales.

"Three of them was in deep water a quarter-mile offshore, but t'other was right into the mouth of Aldridges Cove and he was near as wild as the one inside the Pond. He was running right up to the edge of the shoal ground, blowing high as a steeple, and sending a wash right up onto the shore. I and the young lad hauled clear over to the south side of Fish Rock and took to the lee of the islands. 'Twas a long way out of our course, but I tell you I never liked the looks of that big fellow. Far as I were concerned he could have the whole channel to hisself!"

At their homes in Muddy Hole that Tuesday evening, the Hann brothers knew nothing of this new assault upon the whale. They had spent several quiet hours in her company earlier in the day and had begun to take an almost proprietary interest in her, and even to feel a strengthening sympathy for her predicament.

"What a hard business," Kenneth remembered. " 'Twasn't fitting for a creature the like of she to be barred off. Whales likes company, you see. Was times Doug and me thought she was after looking to we for company . . . Times us'd be gutting cod and that girt big head would come along six, eight feet off the boat, just under water and moving slow and easy as you please, with nary a ripple of a wake. . . . Could see her eye sometimes, looking up at we. We took the habit of saving the herring out of the cod bellies and heaving it overboard when she come nigh. Can't say as she took to it, but 'tis certain she had no use for gurry*! When we'd heave that overboard, she'd go right clear of we until it washed away."

When the Hanns returned to the Pond on Wednes-

---

*The guts of the cod.

day, they noted changes in the whale's behaviour and
appearance.

"She waren't blowing near as high as Tuesday and
she blowed a lot oftener. Couldn't seem to stay down
more'n a few minutes at a time. When she come close
up we see her hide was peppered with white spots the
size of silver dollars. Doug said they was likely bullet
holes but a man could hardly believe she had that
many bullets into her. I thought maybe 'twas where
she'd had barnacles hung onto her, and they got scraped
off.

"Well, sorr, we soon found out the truth of that.
Just afore we left the Pond two of them speedboats
come in from The Reach and six fellows jumped
ashore. They all had them army rifles, and George Old-
ford, he started right in to shooting.

"We hailed them and told them to hold off so we
could get clear of the Pond. George, he yelled down to
we they had orders for to finish off the whale. Put it out
of its misery, like. I told them that was only foolishness. 'You'll shoot somebody yet!' I told them. They
never answered but when we was passing out of the
south gut we heard the lot of them open fire again.

"Doug, he turns to me and says 'tis time somebody
put a stop to it. Trouble was, the Mountie waren't likely
to take a hand unless some of the big folks told him to,
and they was into it as much as anybody . . . People
was needing to use the Pond for the fishing and to fetch
water from The Ha Ha, but 'twas hardly safe to go in
there at all."

Although I still knew nothing about the continuing
attacks on the whale, they were no secret to most people. The firing could be distinctly heard in the eastern
parts of Burgeo and by Thursday a rising indignation
on the part of those who customarily used Aldridges
Pond had brought the matter to a head. Late Thursday
evening some of the Muddy Hole fishermen decided on
a course of action.

During my years in Burgeo I had been called upon
on a number of occasions to speak, or write, on behalf

of individuals and groups who felt themselves incapable of reaching the ears of those in authority. Although a little vague about the real nature of my work as a writer, they assumed I had some kind of influence with those up above.

After supper on Thursday two fishermen from Smalls Island walked into our kitchen bearing gifts of cod tongues and a huge slab of halibut. They sat on the day bed and we talked for a while about the state of the fishery, the weather, and other usual topics of Burgeo life. It was not until they were leaving that the real reason for the visit came out.

"I suppose, Skipper, you knows about the whale?"

"You mean the Finners in The Ha Ha?"

"No, Skipper, I means the one down in Aldridges Pond. Big feller. Been in there quite a time."

"What the devil would a whale be doing in there?" I asked incredulously. "What kind of a whale is it?"

He was vague. "Don't rightly know. Black, like; with a girt big fin. They says it can't get clear . . . Well, goodnight to you, Missus, Skipper."

And with that they vanished.

"Now what do you suppose that's all about?" I asked Claire.

"Who knows? Maybe there's a Pothead caught in Aldridges."

"Maybe. But why did those chaps come all the way up here to tell us about it and then go all evasive when I started asking questions? Something's going on. I think I'll slip over to Sim's and see what he can tell me."

Although I was inclined to agree with Claire that the whale (if it existed at all) would turn out to be a Pothead, I thought there was just a chance it might be a Killer. The remark about the "girt big fin" suggested this and, too, some local dragger men had recently met pods of Killers not far from the Burgeo Islands. Anyway, my curiosity was aroused.

Sim Spencer was alone in his little store, laboriously working up his accounts. Rather reluctantly, it seemed to me, he admitted to having heard something about the Aldridges whale. When I asked him why he hadn't told me before, knowing how interested I was in anything to do with whales, he was embarrassed.

"Well," he said, fumbling for words. "They's been a lot of foolishness . . . a shame what some folks does . . . wouldn't want to bother you with the likes of that . . . but now as you knows, I thinks 'tis just as well."

The implications of this escaped me at the time, but soon became clear enough. The reason I had not been told about the whale was that many of the people were ashamed of what was happening and did not want to talk about it with outsiders; and even after five years I was still something of a newcomer in their midst.

Sim took me to see the Hanns. They were reticent at first but they did describe the whale in fair detail; and I now realized there was an excellent if almost unbelievable chance it might turn out to be one of the great rorquals. Having seen Aldridges Pond in the past, I knew it to be an almost perfect natural aquarium, quite large enough to contain even a Blue Whale in some kind of comfort.

The prospect that, for the first time in history, so far as I knew, it might be possible to come to close quarters with the mystery of one of the mighty lords of Ocean, was wildly exciting. I was in such a hurry to rush home and tell Claire about it that Kenneth Hann's concluding words did not quite sink in.

"They says," he warned, "some fellers been shooting at it. It could get hurted, Skipper, an' they keeps it up."

Probably some damn fool has been taking pot shots at it with a .22, I thought, and put the warning out of mind. As I hurried across Messers bridge in the gathering darkness, my thoughts were fixed on tomorrow, and my imagination was beginning to run away as I contemplated what could happen if the trapped whale indeed turned out to be one of the giants of the seas.

Chapter 11

Early next morning I telephoned Danny Green, a lean, sardonic and highly intelligent man in his middle thirties who had been the high-lining skipper of a dragger but had given that up to become skipper, mate and crew of the little Royal Canadian Mounted Police motor launch. Danny not only knew—and was happy to comment on—everything of importance that happened on the Sou'west Coast, he was also familiar with and interested in whales. What he had to tell me brought my excitement to fever pitch.

"I'm pretty sure 'tis one of the big ones, Farley. Can't say what kind. Haven't seen it meself but it might be a Humpback, a Finner or even a Sulphur." He paused a moment. "What's left of it. The sports have been blasting hell out of it this past week."

As Danny gave me further details of what had been happening, I was at first appalled, then furious.

"Are they bloody well crazy? This is a chance in a million. If that whale lives, Burgeo'll be famous all over the world. *Shooting* at it! What the hell's the matter with the constable?"

Danny explained that our one policeman was a temporary replacement for the regular constable, who was away on leave. The new man, Constable Murdoch, was from New Brunswick. He knew nothing about Burgeo and not much about Newfoundland. He was hesitant to interfere in local matters unless he received an official complaint.

At my request, Danny put him on the phone.

"Whoever's doing that shooting is breaking the game laws, you know," I told him. "It's forbidden to take rifles into the country. Can't you put a stop to it?"

Murdoch was apologetic and cooperative. Not only did he undertake to investigate the shooting, he offered to make a patrol to Aldridges Pond and take me with him. However, Claire and I had already made other arrangements with two Messers fishermen, Curt Bungay and Wash Pink, who fished together in Curt's new boat. They were an oddly assorted pair. Young, and newly married, Curt was one of those people about whom the single adjective, "round," says it all. His crimson-hued face was a perfect circle, with round blue eyes, a round little nose, and a circular mouth. Although he was not fat, his body was a cylinder supported on legs as round and heavy as mill logs. Wash Pink was almost the complete opposite. A much older man, who had known hard times in a distant outport, he was lean, desiccated, and angular. And whereas Curt was a born talker and story teller, Wash seldom opened his mouth except in moments of singular stress.

A few minutes after talking to Murdoch, Claire and I were under way in Curt's longliner. I was dithering between hope that we would find a great whale in the Pond, alive and well, and the possibility that it might have escaped or, even worse, have succumbed to the shooting. Claire kept her usual cool head, as her notes testify:

"It was blowing about 40 miles an hour from the northwest," she wrote, "and I hesitated to go along. But Farley said I would regret it all my life if I didn't. Burgeo being Burgeo, it wouldn't have surprised me if the 'giant whale' had turned out to be a porpoise. It was rough and icy cold crossing Short Reach but we got to Aldridges all right and sidled cautiously through the narrow channel. It was several hours from high tide and there was only five feet of water, which made Curt very nervous for the safety of his brand-new boat.

"We slid into the pretty little Pond under a dash of watery sunlight. It was a beautifully protected natural harbour ringed with rocky cliffs that ran up to the 300-foot crest of Richards Head. Little clumps of dwarfed black spruce clung in the hollows here and there along the shore.

"There was nobody and nothing to be seen except a few gulls soaring high overhead. We looked eagerly for signs of the whale, half expecting it to come charging out of nowhere and send us scurrying for the exit. There was no sign of it and I personally concluded it had left—if it had ever been in the Pond at all.

"I was ready to go below and try to get warm when somebody cried out that they saw something. We all looked and saw a long, black shape that looked like a giant sea-serpent, curving quietly out of the water, and slipping along from head to fin, and then down again and out of sight.

"We just stared, speechless and unbelieving, at this vast monster. Then there was a frenzy of talk.

" 'It's a *whale* of a whale! . . . Must be fifty, sixty feet long! . . . That's no Pothead, not that one . . .'

"Indeed, it was no Pothead but an utterly immense, solitary and lonely monster, trapped, Heaven knew how, in this rocky prison.

"We chugged to the middle of the Pond just as the R.C.M.P. launch entered and headed for us. Farley called to Danny Green and they agreed to anchor the two boats in deep water near the south end of the Pond and stop the engines.

"Then began a long, long watch during which the hours went by like minutes. It was endlessly fascinating to watch the almost serpentine coming and going of this huge beast. It would surface about every four or five minutes as it followed a circular path around and around the Pond. At first the circles took it well away from us but as time passed, and everyone kept perfectly still, the circles narrowed, coming closer and closer to the boats.

"Twice the immense head came lunging out of the water high into the air. It was as big as a small house, glistening black on top and fish-white underneath. Then down would go the nose, and the blow-hole would break surface, and then the long, broad back, looking like the bottom of an overturned ship, would slip into our sight. Finally the fin would appear, at least four feet tall, and then a boiling up of water from the flukes and the whale was gone again.

"Farley identified it as a Fin Whale, the second largest animal ever to live on earth. We could see the marks of bullets—holes and slashes—across the back from the blowhole to the fin. It was just beyond me to even begin to understand the mentality of men who would amuse themselves filling such a majestic creature full of bullets. Why *try* to kill it? There is no mink or fox farm here to use the meat. None of the people would eat it. No, there is no motive of food or profit; only a lust to kill. But then I wonder, is it any different than the killer's lust that makes the mainland sportsmen go out in their big cars to slaughter rabbits or ground-hogs? It just seems so much more terrible to kill a whale!

"We could trace its progress even under water by the smooth, swirling tide its flukes left behind. It appeared to be swimming only about six feet deep and it kept getting closer to us so we began to catch glimpses of it under the surface, its white underparts appearing pale aqua-green against the darker background of deep water.

"The undulations on the surface came closer and closer until the whale was surfacing within twenty feet of the boats. It seemed to deliberately look at us from time to time as if trying to decide whether we were dangerous. Oddly, the thought never crossed my mind that *it* might be dangerous to us. Later on I asked some of the others if they had been afraid of this, the mightiest animal any of us was ever likely to meet in all our lives, and nobody had felt any fear at all. We were too enthralled to be afraid.

"Apparently the whale decided we were not dangerous. It made another sweep and this time that mighty head passed right under the Mountie's boat. They pointed and waved and we stared down too. Along came the head, like a submarine, but much more beautiful, slipping along under us no more than six feet away. Just then Danny shouted: 'Here's his tail! Here's his tail!'

"The tail was just passing under the police launch while the head was under *our* boat, and the two boats were a good seventy feet apart! The flippers, each as long as a dory, showed green beneath us, then the whole unbelievable length of the body flowed under the boat, silently, with just a faint slick swirl of water on the surface from the flukes. It was almost impossible to believe what we were seeing! This incredibly vast being, perhaps eighty tons in weight, so Farley guessed, swimming below us with the ease and smoothness of a salmon.

"Danny told me later the whale could have smashed up both our boats as easily as we would smash a couple of eggs. Considering what people had done to it, why didn't it take revenge? Or is it only mankind that takes revenge?"

Once she accepted the fact that our presence boded her no harm, the whale showed a strange interest in us, almost as if she took pleasure in being close to our two 40-foot boats, whose undersides may have looked faintly whale-like in shape. Not only did she pass directly under us several times but she also passed between the two boats, carefully threading her way between our anchor cables. We had the distinct impression she was lonely—an impression shared by the Hann brothers when she hung close to their small boat. Claire went so far as to suggest the whale was seeking help, but how could we know about that?

I was greatly concerned about the effects of the gunning but, apart from a multitude of bullet holes, none of which showed signs of bleeding, she appeared to be in good health. Her movements were sure and powerful and there was no bloody discoloration in her blow. Because I so much wished to believe it, I concluded that the bullets had done no more than superficial damage and that, with luck, the great animal would be none the worse for her ordeal by fire.

At dusk we reluctantly left the Pond. Our communion with the whale had left all of us half hypnotized. We had almost nothing to say to each other until the R.C.M.P. launch pulled alongside and Constable Murdoch shouted:

·   "There'll be no more shooting. I guarantee you that. Danny and me'll patrol every day from now on, and twice a day if we have to."

Murdoch's words brought me my first definite awareness of a decision which I must already have arrived at below—or perhaps above—the limited levels of conscious thought. As we headed back to Messers, I knew I was committed to the saving of that whale, as passionately as I had ever been committed to anything in my life. I still do not know why I felt such an instantaneous compulsion. Later it was possible to think of a dozen reasons, but these were afterthoughts—not reasons at the time. If I were a mystic, I might explain it by saying I had heard a call, and that may not be such a mad explanation after all. In the light of what ensued, it is not easy to dismiss the possibility that, in some incomprehensible way, alien flesh had reached out to alien flesh . . . cried out for help in a wordless and primordial appeal which could not be refused.

During the run home, my mind was seething with possibilities, with hopes, with fears. Only one thing seemed sure; the whale would need more help than I alone could give. We needed allies, she and I.

As soon as we reached home I called the fish plant manager who, because of his position as "boss," was one of the most influential men in the community. In rather incoherent fashion I tried to convince him of the importance to Burgeo, and to the world at large, of our having one of the great whales in our keeping. A withdrawn, uncommunicative man who seemed to think of everything in managerial terms, he was not easy to persuade; obviously finding it hard to understand why anyone would be much concerned about the life or death of a whale. However, he finally did agree to post notices at the plant asking everyone to leave the animal alone.

He helped me more than he knew, for his very coolness brought me sharply up against the realization that I would have to marshal some convincing practical reasons why the whale should be saved. The most obvious one I could think of was the fact that, to the best of my knowledge, no human being had ever before had the chance to study closely a living rorqual. That chance was here—was now. I was sure the scientific people would recognize its importance, would be as excited as I was, and would come rushing to help. Science had to be alerted.

This was easier said than done. Our telephone link with the outer world consisted of a make-shift system of radio transmitters and micro-wave relays, no part of which was even moderately efficient. Even after a connection had been established with the "outside"—a feat which might take several hours—the chances were excellent that neither party would be able to hear or understand the other. I disliked and distrusted the black machine hanging on our wall so much that I refused to have anything to do with it except in case of dire emergency. But this was an emergency.

The first call I placed was to the Federal Fisheries office in St. John's, Newfoundland. I managed to conduct a shouted conversation with a senior biologist there who explained, kindly, as one would to an enthusiastic but ill-informed child, that his station only concerned itself with fishes . . . and whales were mammals.

Cursing his bureaucratic mind, I spent the next three hours trying to reach the central Fisheries Research station near Montreal. When I eventually got through to its director, I found him sympathetic but no more helpful than his colleague in St. John's. He told me the Department's whale expert was away in the United States studying whale skeletons in museums. It was apparently out of the question that he should be recalled from his research into the bones of

the dead in order to visit Burgeo and study a rorqual
in the living, breathing flesh.

By this time it was late at night and I was beginning to suffer from a growing sense of frustration, coupled with a certain feeling of unreality. Could it be, I
asked myself, that the world of science would prove to
be unmoved by this unique opportunity to gain a
little insight into the life of one of the of the most remarkable animals that ever lived?

In something between panic and despair, I now
called a tried and patient friend, my publisher, Jack
McClelland, in Toronto. Jack resignedly climbed out
of bed to spend the next several hours himself trying to
interest marine biologists all across Canada. Most of
those he reached were outspokenly irate at being roused
in the middle of the night, and none of them apparently gave a damn whether or not Burgeo possessed a captive Fin Whale. One eminent cetologist in
British Columbia listened with cold politeness while
Jack described the situation, then gave him a little
lecture.

Fin Whales, he said, do not eat herring; they sub-
sist entirely on plankton. Thus, even if I did have one
in captivity, I could not possibly feed it. However,
feeding such a captive would be unnecessary in any
case because Fin Whales can easily survive for six
months by assimilating their own blubber. Anyhow,
the point was not germaine, he continued crushingly,
because Fin Whales never come near shore unless they
are dead or dying; therefore, the Burgeo whale, *if* it
was a Fin at all, must be in a dying condition. Since
scientists had already studied a good many dead Fin
Whales, they would not be much interested in studying yet another. He advised Jack to forget the whole
thing.

I was having breakfast early on Saturday morning when Jack phoned back with his doleful news. He
did his best to sound optimistic.

"Look, if you've really got an eighty-ton whale
on your hands, I'll believe you. Evidently not many
people will. But not to worry. You concentrate on

keeping it alive and I'll keep plugging until I find some way to get these silly bastards off their asses."

Saturday had brought another raging blizzard and, since there was no possibility of visiting Aldridges Pond until the weather moderated, I was left to pace the house and consider ways to keep the whale alive. I also had to consider what my ultimate intentions were. Mere success in preserving her life could simply doom her to perpetual captivity in the confines of the Pond, with the likelihood that she would end up as the object of a money-grubbing attempt at exploitation as a tourist attraction. The prospect of saving her life only to deliver her over to the Bander-log was a revolting one. Nevertheless, I was tempted by selfish considerations to equivocate. It was something, after all, to be the nominal possessor of such a fantastic creature. No other human being had ever had a Fin Whale for a "pet." And yet . . . and yet I could not refuse to see that to keep her captive would be to commit another kind of atrocity, almost as cruel as using her for target practise.

Ultimately I could find no way to evade the simple truth. My duty, obligation, purpose—whatever it might be called—did not lie with man; it lay solely with the trapped whale. Whatever I did on her behalf had to be directed toward setting her free. Only so could she be saved.

With this decision made, I had to face the problem of *how* to free her. From local charts, the weather records which I kept in my daily journal, and from the scraps of information I had already picked up from the Hann brothers and other people, I calculated that she had entered the south channel of Aldridges Pond when it carried a maximum depth of about eleven feet. A textbook on the *cetacea* indicated (although there were no concrete data) that an adult Finner, swimming at the surface, carries a draft of just about that much.

The tide tables told me it would be nearly a month before there would again be anything approaching sufficient depth in the channel to float her out to freedom. So that left me with slightly less than a month to plan her release. Considering the difficulties I could expect to encounter in trying to manoeuvre such a titanic beast safely through the narrow channel, a month might be none too long. However, for the moment I felt I could relegate the detailed planning of her escape to the back of my mind while I tried to deal with the immediate difficulties involved in simply keeping her alive.

The first thing, of course, was how to keep her fed. I knew, from my several years of watching Finners at Burgeo, that the whale specialist in British Columbia had been talking through his academic hat when he told Jack that Fin Whales don't eat herring; and I was sure he was just as far at sea when he claimed that a Finner could survive comfortably for six months on its own blubber.

The blubber layer—which is an integral part of a whale's skin—serves only in a secondary role for food storage. Whales developed oil-impregnated tissue primarily as insulation to protect them against the loss of body heat into the hungry conductivity of icy seas. While it is true that a starving whale will, perforce, burn its own blubber oil for fuel, this is a dead-end street in cold northern waters. As the blubber layer thins, ever greater quantities of fuel are required to compensate for the increasing rate of heat loss until, if no other source of food is found, the whale dies of a combination of starvation and exposure.*

Since the temperature of the sea at Burgeo in February hovers around 31° F., just below the freez-

---

*Whales make still another demand on their supplies of stored oil. Although they live surrounded by water, it is *salt water*—which is just as unusable to their metabolism as it is to ours. They have no direct access to *any* source of fresh water. To obtain the vital supplies they need, they must rely on what they can obtain from the body fluids of their prey, supplemented by the chemical breakdown of their oil reserves, which provides fresh water as a by-product. Consequently, if a whale cannot feed, its entire fresh water requirements must be provided from the stored oils, particularly from blubber oil. A whale that is denied food is not only threatened with death by starvation and exposure, it is also doomed to die of thirst.

ing point of fresh water, it seemed all too clear that, without a large and steady supply of food, the trapped whale must perish long before she could be freed.

Food meant herring. But the Hann brothers had told me that Aldridges Pond was rapidly emptying of the little fishes. Either the whale had already decimated the schools or they had fled before the demands of such a gargantuan appetite. It was unlikely that new schools would now enter the Pond of their own free will since there is nothing suicidal about the behavior of a herring school. I would have to find a way to drive them in and pen them there; and at this juncture I had no idea how this could be done.

Equally pressing was the matter of protecting her from the sportsmen. The fact that the Mountie would now try to prevent them from using firearms was no guarantee they would leave her alone. On the contrary, Danny Green had already intimated that they might go to considerable lengths to do her further hurt, if only to spite me for having been instrumental in stopping their target practise. There was a considerable stock of dynamite in Burgeo and it was all too readily accessible. I only hoped this thought would not occur to them but I had no confidence it would not.

Food, protection, and a workable plan to free her at the next spring tide—the *next* spring, not some other—these were her needs, which I had made my problems. It was increasingly clear that I could not do it all alone. I would *have* to find help from the outside.

Before noon on Saturday, I had again returned to the black *djin* of a telephone. This time I decided to appeal directly to the angels. After two hours of wrestling with the intervening demons of air, I managed, by a miracle it seemed, to reach the Minister of Fisheries for Newfoundland on whom, in law at least, the responsibility for the whale rested. I gave him the best pitch I could muster, but my shouted explanations and pleadings brought no tangible result except a hoarseness which was to become chronic in the days ahead. The government of Newfoundland, I was told, had better things to do than concern itself with the preservation of a lone Fin Whale.

Several other calls to both provincial and federal bureaucrats proved equally fruitless. Apparently nobody in authority had any interest in assuming responsibility for the whale or in providing assistance to me in my self-appointed task. I suspect that many of the men I spoke to thought I was a little mad.

Although my previous experiences with Burgeo's resident politicians had given me no grounds to expect help from that quarter, I was driven by my failures elsewhere to turn to our own mayor; but he was absent in St. John's. I thereupon tracked down his deputy, who was the male member of the doctor team. He was a "soft-centred" sort of man, physically graceful in a willowy kind of way, and with a willow's capacity to bend easily before the wind. Although a neutralist by nature, his response to my request for help was distinctly hostile. Not only did he firmly reject the idea that the whale was any of his, or the town council's, business, he was equally emphatic that it was none of mine. His wife, who was also a council member, not only confirmed his judgement but stated her opinion that the Burgeo people had every right to kill the whale by any method they might choose. Furthermore, she said, the carcass could be put to very good use . . . as dog food. (It may have been pure coincidence that the doctors owned those two immense and perpetually hungry Newfoundland dogs).

Darkness had fallen by this time, and the house shook and shuddered in a snow-filled gale. The devils of self-doubt began to stalk me. Perhaps I *was* a little mad—deluded anyway—in thinking I might save the whale. Perhaps the battle was already lost. Perhaps I *had* no business meddling in a tragedy which was essentially a natural one . . . but then I saw again the whale herself, as we had watched her slipping through the green void beneath Curt Bungay's boat. That vision routed the devils instantly. That lost leviathan was one of the last of a disappearing race, and I knew she had to be saved if only because contact with her, though it lasted no more than a few brief weeks, might narrow the immense psychic gap between our two species; might alter, in at least some degree, the remote and awesome image which whales have always projected onto the inner human eye. And if, through this opportunity for intimate contact, that image could be changed enough to let men, to *force* men, to see these secret and mysterious beings with the compassion we have always denied them, it might help bring an end to the relentless slaughter of their kind.

This thought, combined with the effect of the rejections from those whose help I had thus far sought, began to make me fighting mad. I decided that if those who ought to have displayed some interest in the whale refused to do so, I would make them. And, by God, I thought I knew how to do just that.

"Claire," I told my wife, "I'm going to give the story to the press. The *whole* story. About the shooting. There'll be plenty of people who'll react to that. They'll surely make enough fuss, raise enough hell, to force someone out there to act. Burgeo won't like it. It could get damned unpleasant around here. What do you say?"

Claire was very much in love with Burgeo. This was where she had made her first home as a married woman. She understood the delicate nature of our acceptance in the place and she had a woman's shrewd ability to see the possible implications of this decision. Her voice, in reply, seemed very small against the cacophony of the storm.

"If you must . . . oh, Farley, I don't want that whale to die either . . . but you'll be hurting Burgeo . . . the people you like won't understand . . . but I guess . . . I guess it's what you have to do."

The phone rang and I went to it, and it was a reprieve. The operator in Hermitage, her voice barely audible over the babbling static, slowly read me a telegram. It was from Dr. David Sergeant, a biologist with the Federal Department of Fisheries. Sergeant is a maverick and the possessor of a truly open and questing mind. He had taken it on himself to rouse his fellow scientists to action.

> HAVE CONTACTED SEVERAL EMINENT BIOLOGISTS
> NEW ENGLAND THEY VERY EXCITED YOUR WHALE
> SUGGEST YOU BEGIN SYSTEMATIC OBSERVATIONS
> IMMEDIATELY PENDING THEIR ARRIVAL AT EAR-
> LIEST POSSIBLE DATE STOP PHONE CIRCUITS
> BURGEO IMPOSSIBLE BUT WILL TRY AGAIN TOMOR-
> ROW GOOD LUCK.

It was a small enough glimmer of light, but on that dark Saturday it persuaded me that help would come. As we prepared to go to bed, Claire and I were further cheered by a radio forecast which promised an end to the storm and a fine Sunday coming.

If I had guessed what that Sunday would bring, I think I would have prayed for a hurricane.

Chapter 12

Burgeo winter weather often seemed to consist of six days of storm followed by a seventh when all was forgiven, and the seventh day was almost always a Sunday. I once discussed this interesting phenomenon with the Anglican minister but he decently refused to take any credit for it.

Sunday, January 29th, was no exception. It almost seemed as if spring had come. The sun flared in a cloudless sky; there was not a breath of wind; the sea was still and the temperature soared.

Early in the morning Onie Stickland and I went off to Aldridges in his dory. We took grub and a tea kettle since I expected to spend the entire day observing the whale and noting her behaviour for the record. I hoped Onie and I would be alone with her, but there were already a number of boats moored to the rocks at the outer end of the channel when we arrived, and two or three dozen people were clustered on the ridge overlooking the Pond. I saw with relief that nobody was carrying a rifle.

We joined the watchers, among whom were several fishermen I knew, and found them seemingly content just to stand and watch the slow, steady circling of the whale. I used the opportunity to spread some propaganda about Burgeo's good luck in being host to such a beast, and how its continued well-being would help in drawing the attention of faraway government officials to a community which had been resolutely neglected for many years.

The men listened politely but they were sceptical. It was hard for them to believe that anyone outside Burgeo would be much interested in a whale. Nevertheless, there did seem to be a feeling that the whale should not be further tormented.

"They's no call for that sort of foolishness," said Harvey Ingram, a lanky, sharp-featured fisherman, originally from Red Island. "Lave it be, says I. 'Tis doing harm to none."

Some of the others nodded in agreement and I began to wonder whether—if no help came from outside—it might be possible to rouse sufficient interest in the whale, yes, and sympathy for her, so we could take care of her ourselves.

"Poor creature has trouble enough," said one of the men who fished The Ha Ha. But then he took me down again by adding:

"Pond was full of herring first day she come in. Now we sees hardly none at all. When we first see the whale, 'twas some fat, some sleek. Now it looks poorly. Getting razor-backed, I'd say."

We were interrupted by the arrival, in a flurry of spray and whining power, of a big outboard speedboat, purchased through the catalogue by one of the young men who spent their summers on the Great Lakes freighters. He was accompanied by several of his pals, all of them sporting colourful nylon windbreakers of the sort that are almost uniforms for the habitués of small-town poolrooms on the mainland. They came ashore, but stood apart from our soberly dressed group, talking among themselves in tones deliberately pitched high enough to reach our ears.

"We'd a had it kilt by now," said one narrow-faced youth, with a sidelong glance in my direction, "only for someone putting the Mountie onto we!"

"And that's the truth!" replied one of his companions. "Them people from away better 'tend their own business. Got no call to interfere with we." He spat in the snow to emphasize his remark.

"What we standing here for?" another asked loudly. "We's not afeared of any goddamn whale. Let's take a run onto the Pond. Might be some sport into it yet."

They ambled back to their powerboat and when the youths had clambered aboard, one of the men standing near me said quietly:

"Don't pay no heed, skipper. They's muck floats up in every place. Floats to the top and stinks, but don't mean nothin'."

It was kindly said, and I appreciated it.

By this time a steady stream of boats was converging on the Pond from Short Reach, The Harbour, and from further west. There were power dories, skiffs, longliners and even a few rowboats with youngsters at the oars. Burgeo was making the most of the fine weather to come and see its whale.

The majority of the newcomers seemed content to moor their boats with the growing armada out in the entrance cove, but several came through the channel into the Pond, following the lead of the mail-order speedboat. At first the boats which entered the Pond kept close to shore, leaving the open water to the whale. Their occupants were obviously awed by the immense bulk of the creature, and were timid about approaching anywhere near her. But by noon, by which time some thirty boats, bearing at least a hundred people, had arrived, the mood began to change.

There was now a big crowd around the south and southwest shore of the Pond. In full awareness of this audience, and fortified by lots of beer, a number of young men (and some not so young) now felt ready to show their mettle. The powerful boat which had been the first to enter suddenly accelerated to full speed and roared directly across the Pond only a few yards behind the whale as she submerged. Some of the people standing along the shore raised a kind of ragged cheer, and within minutes the atmosphere had completely—and frighteningly—altered.

More and more boats started up their engines and nosed into the Pond. Five or six of the fastest left the security of the shores and darted out into the middle. The reverberation of many engines began to merge into a sustained roar, a baleful and ferocious sound, intensified by the echoes from the surrounding cliffs. The leading powerboat became more daring and snarled across the whale's wake at close to twenty knots, dragging a high rooster-tail of spray.

The whale was now no longer moving leisurely in great circles, coming up to breathe at intervals of five or ten minutes. She had begun to swim much faster and more erratically as she attempted to avoid the several boats which were chivvying her. The swirls of water from her flukes became much more agitated as she veered sharply from side to side. She was no longer able to clear her lungs with the usual two or three blows after every dive, but barely had time to suck in a single breath before being driven down again. Her hurried surfacings consequently became more and more frequent even as the sportsmen, gathering courage because the whale showed no sign of retaliation, grew braver and braver. Two of the fastest boats began to circle her at full throttle, like a pair of malevolent water beetles.

Meanwhile, something rather terrible was taking place in the emotions of many of the watchers ringing the Pond. The mood of passive curiosity had dissipated, to be replaced by one of hungry anticipation. Looking into the faces around me, I recognized the same avid air of expectation which contorts the faces of a prizefight audience into primal masks.

At this juncture the blue hull of the R.C.M.P. launch appeared in the entrance cove. Onie and I jumped aboard the dory and intercepted her. I pleaded with Constable Murdoch for help.

"Some of these people have gone wild! They're going to drive the whale ashore if they don't drown her first. You have to put a stop to it . . . order them out of the Pond!"

The constable shook his head apologetically.

"Sorry. I can't do that. They aren't breaking any law, you know. I can't do anything unless the local authorities ask me to. But we'll take the launch inside and anchor in the middle of the Pond. Maybe that'll discourage them a bit."

He was a nice young man but out of his element and determined not to do anything which wasn't "in the book." He was well within his rights; and I certainly overstepped mine when, in my distress, I intimated that he was acting like a coward. He made no reply, but quietly told Danny to take the police boat in.

Onie and I followed them through the channel, then we turned along the southwest shore where I hailed several men in boats, pleading with them to leave the whale alone. Some made no response. One of them, a middle-aged merchant, gave me a derisive grin and deliberately accelerated his engine to drown out my voice. Even the elder fishermen standing on shore now seemed more embarrassed by my attitude than sympathetic. I was slow to realize it but the people gathered at Aldridges Pond had sensed that a moment of high drama was approaching and, if it was to be a tragic drama, so much the better.

Having discovered that there was nothing to fear either from the whale or from the police, the speedboat sportsmen began to make concerted efforts to herd the great beast into the shallow easterly portion of the Pond. Three boats succeeded in cornering her in a small bight, and when she turned violently to avoid them, she grounded for half her length on a shelf of rock.

There followed a stupendous flurry of white water as her immense flukes lifted clear and beat upon the surface. She reared forward, raising her whole head into view, then turned on her side so that one huge flipper pointed skyward. I had my binoculars on her and for a moment could see all of her lower belly, and the certain proof that she was female. Then slowly and, it seemed, painfully, she rolled clear of the rock.

As she slid free, there was a hubbub from the crowd on shore, a sound amounting almost to a roar, that was audible even over the snarl of engines. It held a note of insensate fury that seemed to inflame the boatmen to even more vicious attacks upon the now panic-stricken whale.

Making no attempt to submerge, she fled straight across the Pond in the direction of the eastern shallows where there were, at that moment, no boats or people. The speedboats raced close beside her, preventing her from changing course. She seemed to make a supreme effort to outrun them and then, with horrifying suddenness, she hit the muddy shoals and drove over them until she was aground for her whole length.

The Pond erupted in pandemonium. Running and yelling people leapt into boats of all shapes and sizes and these began converging on the stranded animal. I recognized the doctor team—the deputy mayor of Burgeo and his councillor wife—aboard one small longliner. I told Onie to lay the dory alongside them and I scrambled over the longliner's rail while she was still under way. By this time I was so enraged as to be almost inarticulate. Furiously I *ordered* the deputy mayor to tell the constable to clear the Pond.

He was a man with a very small endowment of personal dignity. I had outraged what dignity he did possess. He pursed his soft, red lips and replied:

"What would be the use of that? The whale is going to die anyway. Why should I interfere?" He turned his back and busied himself recording the whale's "last moments" with his expensive movie camera.

The exchange had been overheard, for the boats were now packed tightly into the cul-de-sac and people were scrambling from boat to boat, or along the shore itself, to gain a better view. There was a murmur of approval for the doctor and then someone yelled, gloatingly:

"Dat whale is finished, byes! It be ashore for certain now! Good riddance is what I says!"

Indeed, the whale's case looked hopeless. She was aground in less than twelve feet of water; and the whole incredible length of her, from the small of the tail almost to her nose, was exposed to view. The tide was on the ebb and if she remained where she was for even as little as half an hour, she would be doomed to die where she lay. Yet she was not struggling. Now that no boats were tormenting her, she seemed to ignore the human beings who fringed the shore not twenty feet away. I had the sickening conviction that she had given up; that the struggle for survival had become too much.

My anguish was so profound that when I saw three men step out into the shallows and begin heaving rocks at her half-submerged head, I went berserk. Scrambling to the top of the longliner's deckhouse, I screamed imprecations at them. Faces turned toward me and, having temporarily focused attention on myself, I launched into a wild tirade.

This was a *female* whale, I cried. She might be and probably was pregnant. This attack on her was a monstrous, despicable act of cruelty. If, I threatened, everyone did not instantly get to hell out of Aldridges Pond and leave the whale be, I would make it my business to blacken Burgeo's name from one end of Canada to the other.

Calming down a little, I went on to promise that if the whale survived she would make Burgeo famous. "You'll get your damned highway!" I remember yelling. "Television and all the rest of it . . ." God knows what else I might have said or promised if the whale had not herself intervened.

Somebody shouted in surprise; and we all looked. She was moving.

She was turning—infinitely slowly—sculling with her flippers and gently agitating her flukes. We Lilliputians watched silent and incredulous as the vast Gulliver inched around until she was facing out into the Pond. Then slowly, slowly, almost imperceptibly, she dirfted off the shoals and slid from sight beneath the glittering surface.

I now realize that she had not been in danger of stranding herself permanently. On the contrary, she had taken the one course open to her and had deliberately sought out the shallows where she could quite literally catch her breath, free from the harassment of the motorboats. But, at the time, her escape from what appeared to be mortal danger almost seemed to savour of the miraculous. Also, as if by another miracle, it radically altered the attitude of the crowd, suddenly subduing the mood of feverish excitement. People began to climb quietly back into their boats. One by one the boats moved off toward the south channel, and within twenty minutes Aldridges Pond was empty of all human beings except Onie and me.

It was an extraordinary exodus. Nobody seemed to be speaking to anybody else . . . and not one word was said to me. Some people averted their eyes as they passed our dory. I do not think this was because of any guilt they may have felt—and many of them *did* feel guilty—it was because *I* had shamed *them,* as a group, as a community, as a people . . . and had done so publicly. The stranger in their midst had spoken his heart and displayed his rage and scorn. We could no longer pretend we understood each other. We had become strangers, one to the other.

My journal notes, written late that night, reflect my bewilderment and my sense of loss.

". . . they are essentially good people. I know that, but what sickens me is their simple failure to resist the impulse of savagery . . . they seem to be just as capable of being utterly loathesome as the bastards from the cities with their high-powered rifles and telescopic sights and their mindless compulsion to slaughter everything alive, from squirrels to elephants . . . I admired them so much because I saw them as a natural people, living in at least some degree of harmony with the natural world. Now they seem nauseatingly anxious to renounce all that and throw themselves into the stinking quagmire of our society which has perverted everything natural within itself, and is now busy destroying everything natural outside itself. How can they be so bloody stupid? How could *I* have been so bloody stupid?"

Bitter words . . . bitter, and unfair; but I had lost

my capacity for objectivity and was ruled, now, by irrational emotions. I was no longer willing, or perhaps not able, to understand the people of Burgeo; to comprehend them as they really were, as men and women who were also victims of forces and circumstances of whose effects they remained unconscious. I had withdrawn my compassion from them, in hurt and ignorance. Now I bestowed it all upon the whale.

Chapter 13

As with so many aspects of the life of the Fin Whale, so it is with their intimate and personal relationships —we know almost nothing about the subject. We have never seen them in the act of making love. No man has ever witnessed the birth of one. We do not even know with any certainty how long the gestation period is; how often a female gives birth; how old she is at sexual maturity; or even how she manages to suckle her young under water.

Examination of foetuses taken from dead Finners suggests that the young are born in early spring, perhaps in March or April. At any rate, this is presumed to be the case amongst Finners who live in the North Atlantic. Because late-term foetuses have only a very thin blubber layer, and therefore not much insulation, some biologists believe the young must be born in warm southern waters, perhaps in the mysterious region near the Sargasso Sea. Other cetologists are equally positive that, because of the quantity and extreme richness of the mother's milk (it is ten times richer than that of a Jersey cow), baby Fins can produce enough internal heat to enable them to survive even in far northern waters. These men suggest that the young are born near the edge of the arctic ice pack. But the fact is that nobody knows.

Again basing their conclusions on examination of dead foetuses, biologists surmise that the gestation period is between ten and twelve months and they think that, at birth, young Finners must be eighteen to twenty feet in length and weigh nearly two tons! The growth rate after birth seems to be equally fantastic. Avidly guzzling at its mother's twin breasts, the young whale is thought to grow to a length of about forty-five feet, and a weight of perhaps twenty tons, during a nursing period between six to eight months. After that its growth slows. It apparently takes at least six and probably as many as eight years for the youngsters to reach puberty, by which time the males—always somewhat smaller than their mates—may be sixty feet long and the females about sixty-five. Although sexual maturity seems to come relatively early, there is new evidence suggesting that Finners are not fully grown until about the age of thirty, by which time a female may be seventy-five feet long and weigh as much as ninety tons. Until recently, science thought the Finner's life span was rather brief—perhaps twenty or thirty years. But during the last few years a method of ageing baleen whales, by counting the number of concentric growth rings in their horny ear plugs, has been developed and it now seems certain that, left unmolested, a Blue or a Fin can expect to surpass three score years and ten with ease. In fact the baleen whales may be the most long-lived of all mammals, including man.* And since they are preyed upon by no natural enemies in adulthood, except, of course, for us, and appear to be singularly free from fatal diseases, they are probably one of the very few non-human forms of life that nature would permit to die of old age, if man did not intercede.

---

*Probably we will never know what the "normal" life span of any of the great whales really was. Because the oldest were also the biggest, they were prime targets of harpooners in the modern catchers. So thorough was the hunt for the big ones that, since 1950, *almost no fully mature rorquals of any species have been taken.* Few, if any, are left alive. However, scientists recently examined the ear plug of one of the last really large Finners, which had been killed half a century ago (the plug had been kept in preservative). They estimated its age as between eighty and ninety years when a harpoon ended its life.

It is my belief that, until they are sexually mature, the young remain in company with their parents as part of the family group—the pod. Unhappily, it is during this youthful period of relative inexperience that they are most vulnerable to the whalers, and catch records show that at least half the Fin Whales killed in recent years never even had a chance to mate and so help perpetuate their kind. Today, as we hunt them toward extinction, the family pods are almost always small; however, records from earlier whaling days show that Fin families often numbered as many as eight individuals.

We know nothing about their courtship or even how the young whales find their mates; but we can guess that both events used to take place during the periodic assemblies when all the families which occupied a given portion of the ocean came together for a while. Tales of such concentrations are common in the old records although none has been reported in North Atlantic waters during the past forty years.

Today the few remaining Fin Whale families are so widely scattered that a young Finner may have to wait many years before encountering a potential mate. This is the more deeply tragic because Finners seem to be strictly monogamous. There is nothing to indicate that a sexually mature daughter ever produces young while she remains in the family pod, or that a widowed female will mate again except with an unattached male. Polygamy, which is the rule among Sperm Whales, has helped that nation to partly hold its own against our depredations. But the practice of monogamy among the Finners may prove to be a luxury their decimated species cannot afford.

The love-making of the Fin may always remain a secret; and I for one will not regret it; let them keep their tender intimacies well hidden. However, this we do know: the bonds between a mated pair are of legendary tenacity; and if this be not love, then love is nothing. Whalers have long been aware of this, and have bloodily profited from the knowledge. They knew that if they could harpoon the female in a pod, her mate would remain by her, so completely reckless of the risk that he all too often joined her in death.

The reverse is not always true. A female will abandon her endangered mate if she is pregnant or has a calf in tow. However, if she is not driven by the need to protect the next generation, she too will often remain with a dying mate until the bombs explode deep in her own vitals.

I knew one old Scots gunner who in his day had killed more than two thousand whales but who had never overcome his revulsion at striking a female of the rorqual tribe.

"We never wanted to know too much about them," he explained. "It was too much like murder as it was. I think if I'd had the Celtic gift of 'sight' and could have looked into the minds of those beasts, I'd have had to give up the sea and go ashore for good. There're times when too much knowledge can stand in a man's way."

The disastrous events of Sunday, coupled into the bargain with the discovery that the whale was a female who might very well be pregnant, made it even more urgent that I obtain help. I decided I would have to follow through on my earlier decision and at 10 o'clock on Monday I sent the following telegram to the Canadian Press head office in Toronto:

SEVENTY-FOOT WHALE WEIGHING ABOUT EIGHTY TONS TRAPPED LARGE SALTWATER POND BURGEO SINCE JANUARY 21 STOP POND FORMS NATURAL AQUARIUM HALF BY HALF MILE DIMENSIONS ALLOWING WHALE CONSIDERABLE FREEDOM MOVEMENT STOP DURING FIRST FIVE DAYS LOCAL SPORTS USED WHALE AS TARGET HIGH-POWERED RIFLES AND STILL CONTINUE HARASSMENT WITH SPEEDBOATS STOP HAVE PREVAILED RCMP HALT SHOOTING BUT FEARFUL OTHER DANGERS STOP THIS IS FIRST GREAT WHALE EVER REPORTED SIMILAR CIRCUMSTANCES PROVIDING ABSOLUTELY UNEQUALLED POSSIBILITIES FOR STUDY BUT URGENTLY REQUIRE ASSISTANCE PROTECT ANIMAL

AND ORGANIZE FEEDING PROGRAM STOP WHALE
RAPIDLY LOSING WEIGHT OTHERWISE APPEARS
GOOD CONDITION IS TOLERANT HUMAN BEINGS
DESPITE PERSECUTION STOP FOR FURTHER DE-
TAILS PHONE ME BURGEO.

I was hardly sanguine enough to believe this sparse
account would set the media world afire; I only hoped
it would be of enough interest to the press and radio
people so they would call back for more information,
out of which they might make a story that would stir
someone in the outside world to action. Consequently, Claire and I were flabbergasted when we
switched on the Canadian Broadcasting Corporation's
noon report of world and national news and heard
my account of the trapped whale already being head-
lined.

Luck, which until then had run so heavily against
her, seemed to have veered in the whale's favour,
and for a reason which we could not have imagined.

Because of our long absence away from Canada,
we had not been aware of a story which had been running for many weeks about a pod of White Whales
(relatively small, porpoise-like, toothed whales) which
became trapped by an early freeze-up in a long inlet on
the arctic coast near Inuvik, a small community near
the mouth of the Mackenzie River. Unable to escape
to the open sea beneath fifty miles of new ice, the
seventeen whales in the pod were at first able to keep
a breathing space open by their own efforts. But as the
weather grew colder and their little patch of open
water inexorably contracted, their situation became crit-
ical.

The plight of these little whales roused great interest in Inuvik, and a committee was formed to try
and save them. By early January, the ice had closed
in so completely that the whales only had a strip of
open water forty feet long by twenty wide in which
they could surface. The Inuvik committee had flown
in power saws with which to keep the breathing hole
open and the battle to save the White Whales had become a national cause.

On that same balmy Sunday when the trapped Finner at Burgeo, some 6000 miles to the southeast, was being harried by sportsmen in motorboats, a full-scale arctic blizzard was sweeping into the Inuvik region, sending the temperature plunging to 40° below zero and preventing anyone from reaching the inlet where the White Whales were trapped. On Monday, even as my telegram was clicking into the Canadian Press office in Toronto, a wire was on its way to that office from Inuvik, bearing the news that the breathing hole had frozen over during the night and the White Whales had perished.

The tragic conclusion to the Inuvik story and the breaking of the Burgeo story appeared on the desks of news editors almost simultaneously and the editors were not slow to make the transposition from one story to the other. As Monday wore on, the telephone circuits linking Burgeo with the outer world began to overload to the point of collapse with calls from radio stations, newspapers and wire services, all seeking amplification of the brief C.B.C. report. The young woman who operated the radio relay station at Hermitage had to perform superhuman feats to keep communications open. I never did know her name; but later on, after a particularly hectic night during which she exhausted herself on the whale's behalf, she so far departed from protocol as to call me personally and, in a voice tearful with fatigue, assure me she would continue to do everything she could to keep the line functioning "so that poor beast has got a chance."

The media demands prevented me from visiting the whale on Monday, but Danny Green kept me posted. "She's swimming as smart as ever," he reported by phone. "Right quiet, too, and blowing stronger than yesterday. The quare thing though is t'other whale. He was right off the cove all the time we was to the Pond, and the Hann boys says he was there every toime they went in and out. Constable Murdoch and me watched for a good while, and here's the quarest thing; both them whales was spouting right together; and both was sounding together, though they was half a mile apart and never could have seen each other. Maybe 'tis foolish, but I believes they's a pair and they talks somehow. You say the one inside's a she? Well, bye, *I* says the one outside's a he!"

Although the noises made by the smaller, toothed whales, as they are used for sound ranging and echo location, have been studied, we have barely begun to investigate how these complicated sounds are used for communication. That they *are* so used is not in doubt. The studies on dolphins of Dr. John C. Lilly, while not as conclusive as some orthodox scientists would wish, have made this point. Lilly, and those who have worked with him, have shown that dolphins possess intelligence—alien to ours as it must be—which is nevertheless worthy of comparison with ours; and that these relatively (as compared to the rorquals) primitive little whales can not only exchange complex information but can also transmit to one another rich emotional feelings to a degree unsurpassed by any non-human animal of which we have any knowledge.*

As yet we can only guess at the communicatory capabilities of the rorquals and other great whales. Until two decades ago it was actually believed by science that most, if not all, the baleen whales were completely dumb! Although they had been hunted for centuries, apparently no man had ever heard one of them utter a single sound. However, the use of supersensitive hydrophones (designed to eavesdrop on enemy submarines) has recently resulted in the astonishing discovery that the rorquals are amongst the most "talkative" of living beings. The range, complexity, and frequency of their outpourings is so great that the few scientists who have studied rorqual sounds admit to being completely baffled when it comes to interpreting, or understanding, their modes, purposes or meanings. Some of the weirdly melodious sequences may very well be music in the highest sense of the

*See *Man and Dolphin,* 1961; and *The Mind of the Dolphin,* 1967; by Dr. John C. Lilly.

word. Other incredibly complex combinations of high-frequency clicks and whistling sounds are uncommonly like high-speed communication codes. It may be a long time before we crack these codes, if indeed we ever do. In the meantime, anybody with an open mind who listens to underwater recordings of the Humpback Whale, for instance, will find it extraordinarily difficult to resist the conclusion that these rorquals can and *do* communicate with each other on levels of content and efficiency which we may have reason to envy.* As to *what* they have to say to each other, we have only the faintest of clues. Still, we can be reasonably confident that they are not just talking for the sake of hearing the sound of their own voices. They seem far too intelligent for that.

Whale talk needs no electronic aids in order to span great distances. Water is a much better conductor of sound than is air, and even with *our* relatively inefficient hearing, we can listen-in to Fin Whales talking underwater, with their low-frequency ranges, at distances of up to thirty-five miles! There is strong reason to believe that some of the great whales can communicate with each other when they are many hundreds of miles apart, and I know one scientist who suspects that whale "talk" may be transmitted right across ocean basins through peculiar "carrier corridors" of water deep in the oceans. The exotic properties of these corridors have only recently been discovered by men, and they are now being exploited for military purposes so that not much has been said about the matter publicly. My friend, who occasionally works for the U.S. Navy, is convinced that whales know about these global communication channels, and may use them for "long distance" calls, free of any tolls.

The "Guardian," as we christened him, was still at his post Tuesday morning when, despite snow squalls

---

*Songs of the Humpback Whale, recorded by Dr. Roger S. Payne, is available on Capital Records.

and a wicked wind, I fled from the media monster I had unleashed, to the gentler company of the quiet monster in the Pond. Onie took me in his dory and we were still some distance from the entrance cove when we saw the Guardian send his "spray," as Onie called it, high into the murky air.

We cut the engine and let the dory drift down toward the whale who was behaving in a most unusual way. He was circling at speed in a space not more than two hundred yards in diameter at the very mouth of the cove, and rising to blow at intervals of only a minute or two. He appeared to ignore our presence, lingering briefly on the surface even when we had drifted to within fifty feet of him. It was then I heard the voice of the Fin Whale once more, and this time under circumstances which left no doubt about its source.

Again it was something felt as well as heard: a deep, vibrant sound such as might perhaps be simulated by a bass organ pipe heard from a distance on a foggy night. It was a deeply disturbing sound, a kind of eery ventriloquism out of another world and utterly foreign to anything Onie and I were familiar with. Hoping for a repetition, we waited silent in the dory until the little boat had drifted well past the entrance, but the Guardian whale had sounded and we heard the voice no more. I told Onie to spin the flywheel and we nosed into the Pond.

We had no sooner cleared the channel than the lady whale spouted close by . . . spouted and instantly submerged as a big, white speedboat roared down upon her at such a clip that the four men in it did not even see us until they had almost swamped us with their wash. They circled and throttled down, and I recognized some of the enemy from Sunday.

I jumped to my feet. "You get the hell out of here!" I yelled furiously. "Get out now and don't come back!"

The driver idled his outboard and grinned defiantly.

"I suppose you can make we go?" he challenged.

I bluffed. "The Mountie sure and hell can! Premier Smallwood has taken over this whale. It's government property now."

In the Newfoundland of 1967 there was but one God, and Joey Smallwood was his Prophet. Although Smallwood had as yet betrayed no interest in the whale, I did not hesitate to take his name in vain. It was the only threat which could have had any effect upon these men. There was some muttering, but in the end the speedboat and its occupants departed.

Onie and I went ashore and settled ourselves in the lee of a commanding rock. With the return of peace, the whale resumed her circling routine but at first she surfaced only at the north end of the Pond, the farthest point away from our moored dory. An hour passed before she came up close enough so that we could see, with horror, a great slash some three or four feet long across her back and just foward of her fin. The white blubber was laid bare to a depth of several inches. When the sportsmen in the white speedboat went back to Burgeo, they described to some of their friends how they had bravely planed their boat at high speed over the whales' back as she was submerging.

"Bust a sheer pin," one of them bragged, "but we cut a Jesusly big hole into her!"

Despite the ugly appearance of the whale's new injury, it did not seem to distress her or to interfere with her activities which, on this day, included something I had heard about from Uncle Art and others, but had not seen before. Shortly before high tide she suddenly stopped her slow patrolling and dashed, swiftly and purposefully, toward the middle of the Pond where she began circling so close to the surface that the water boils from her flukes made a continuous pattern of interlocked rings. Then she again altered course. There was a flash of greenish-white light reflected from her undersides, followed by a swirl of water and rising bubbles which signified that she had opened her cavernous mouth.

"She got a smell of herring!" Onie cried excitedly.

"Tides up now. A bit of a scull must have slipped in through the gut and she took a run at they!"

It must have been a *very* little school, for we were perched where we could see the bottom of the channel and could hardly have avoided noticing even a minor run of the little fishes. Nor did the whale make any more attempts to feed that day.

"I t'inks she'll starve on what herring comes into the Pond of its own," was Onie's opinion. "She must be some hungry. Too bad she wouldn't glutch down ary o' them young connors in the speedboat. Put them to a mite of good, it would."

We remained at the Pond until dark when, almost frozen, we headed home. Apart from the incident with the speedboat, it had been a quiet day at Aldridges . . . but things had been quite different at Messers. I had sown the wind and poor Claire had reaped the whirlwind. She was numb from listening to more than thirty phone calls and telegrams from newspapers, wire services, radio stations, and even one from the Premier of Newfoundland.

DELIGHTED TO BE ABLE TO TELL YOU MY COL-
LEAGUES HAVE ACCORDED MY REQUEST THAT WE
PAY UP TO ONE THOUSAND DOLLARS TO THE FISH-
ERMEN OF BURGEO TO ENABLE THEM TO SUPPLY
HERRING FOR YOUR WHALE STOP WOULD YOU UN-
DERTAKE TO ORGANIZE CARE AND FEEDING OF
YOUR CATCH STOP KINDEST REGARDS.

J R SMALLWOOD

As I was reading the telegram, Claire said, "I had a phone call from a St. John's reporter. He told me the whale story was national and even international news already, and Joey was going to climb on the bandwagon. He also said to tell you not to spend any of that thousand until you had it in your hands."

The day had also brought photographer Bob Brooks from the *Toronto Star*. Brooks had arrived late in the afternoon aboard a chartered ski plane that had landed him two miles inland on the ice of a small lake, from whence he had to slog his way through knee-deep snow to the shores of Short Reach. Luckily for him a passing fisherman saw him there and ferried him across to Burgeo just before he froze.

He was still indignant about the isolation of Burgeo when I arrived home to find him thawing out before our stove.

"Hell," he said feelingly, "it's easier to get to Baffin Island than here. *What* a place you chose to live!"

Not all the calls Claire answered that day were from away. One of them came from the female member of the doctor team. She was furious because I had dared tell the outside world about the whale—as if I were some kind of informer. But that hardly mattered because Smallwood's message had got around and it was beginning to dawn on our businessmen, even on our politicians, that a real, live whale right here in Burgeo had publicity value. From being no more than a massive curiosity, fit only to provide a target for the local sports, she now began to look like something different—like money.

My first intimation of her new status came late that evening. Claire and I were discussing how we should handle contributions of money which, we had been told by several reporters, were already being collected to help feed the whale.

"Why not ask the Sou'westers Club?" Claire suggested. "They're a service club. They ought to be glad to do it."

I had my doubts. Nevertheless, I called one of the club's officers. To my surprise, he not only listened with interest—he was positively enthusiastic.

"Too bad you told them about the gunning," he began. "But that's no great matter. The whale has sure put Burgeo on the map. Even got Joey interested, I hear. Could be the best thing ever happened here. Sure, we'll be glad to look after the money. And anything else needs doing, you just call on us."

I rang off feeling much relieved. The backlash of resentment which I had feared would result from my press release did not seem to have materialized.

My last task that Monday night was to visit the

Hanns to ask their opinion about whether herring were entering the Pond in any quantity. They thought that few were entering and that those few would soon be driven out again by the whale's presence. So I arranged for them to make a trip to the Pond, just after high tide that night, and bar off the entrance with a length of fine-mesh net which would prevent any herring which might have entered a high water from escaping again. This was only an interim measure, of course. I knew we must soon find a really effective way of getting feed to her—in quantity.

Meanwhile, Claire had been busy painting a large sign to be put up at the entrance to the channel. Maybe she stretched a point in the wording but, after all, Joey Smallwood *had* publicly appointed himself patron of the whale, and he *was* the government of Newfoundland.

WARNING

**THIS WHALE
MUST NOT BE
TORMENTED**

Aldridges Pond Is
Closed To All Boats
Without Permission.

By order
Govt. of Newfoundland

Chapter 14

February opened with what promised to be a superb day; frosty and windless, with an illimitable sky sunfaded to pale azure. Onie Stickland was on hand even before we finished breakfast, as anxious as I myself was to see how the whale was making out.

Unmarried, and living with his nephew's family, Onie seemed to lead a contented life. Everyone liked him, and he was friendly with everyone, yet there was an elusive quality about him—something untouchable, not to be fathomed. His long, melancholy face concealed its own mystery. He was a background man, a slight, shadowed figure who could pass unnoticed in any crowd.

If he had a confidante it was his nephew's black waterdog, Rover. Sometimes man and dog would disappear together for hours on end. Once when I was scunning for whales I happened to turn my binoculars on a strip of beach some distance away, and there was Onie sitting on a balk of driftwood. His hand was on the old dog's burly head; hs body strained forward in a listening attitude, immobile, waiting or watching for something known only to himself. For the rest, his interest in the Burgeo scene seemed passive, almost a little vague . . . until the coming of the whale.

From the first, Onie was strangely drawn to her. In his unobtrusive way he would make himself and his dory available any time I wanted to visit the Pond, and he never asked for pay for his services. He would sit in his dory, or on the shore rocks, watching the stately progress of the whale as she circled round and round. He seldom took his eyes off her. I was not fully aware of the depth of his absorption until one evening when we were leaving the Pond. Ahead of us the Guardian surfaced and blew and, like an echo, there came the whooshing exhalation of the lady whale behind us. Onie stopped poling the dory through the channel and quietly, but with great intensity, said:

"*Everyt*'ing ought to be free to go where it wants!"

As if he had given away too much, he turned his back upon me and quickly bent to the flywheel of the old engine.

Those few words stuck in my mind and led me to inquire more deeply into the nature of Onie's apparently contented life. Only then did I learn that, all through youth and early manhood, he had supported his ailing parents and, after they died, had supported and stood by a crippled sister. His life, which I had tended to think about, romantically, as that of an independent and solitary-minded fisherman, spent of his own choice in the security of a quiet outport, had, in bitter truth, been a life lived out in prison—trapped—for he, too, had dreamed. All his life Onie Stickland had hungered to go to sea—not fishing and not coasting, but *deep* sea; to roam the oceans of the world.

If I had known this earlier, I would not have been surprised that he took the plight of the whale so much to heart. He understood, and he pitied her who had once been free and now was free no more.

Accompanied by Bob Brooks (heavily garlanded with cameras), we arrived at the Pond to find it deserted by mankind. We took time to erect the new notice board at the mouth of the channel and to refasten the net which the Hanns had used to bar off the entrance, and which had unaccountably come adrift. Then we crossed the intervening ridge.

The Pond was absolutely calm, faultlessly mirroring the surrounding hills and cliffs. Sombre it may have been, but there were many subtle colours in the rocks, and in their reflections, floating faithfully upon a sheet of water that seemed to burn in its very depths with a still, blue flame.

The whale rose into view almost at once, blowing near an islet on the northern shore. I set myself up with binoculars, notebook and stop-watch to observe and record her actions, while Brooks scrambled off among the cliffs to look for camera angles. There was a marvelously dreamy quality about the morning. The movements of the whale were very rhythmic. She swam clockwise—with the sun, as sailors say—remaining submerged many minutes at a time before surfacing to blow once or twice, and then sinking slowly out of sight again. Because of the calmness and icy clarity of the water, I could sometimes see her whole body shimmering beneath the surface.

She was not fishing, for there was probably nothing to catch. Nevertheless, she kept up her steady, fluid progress round and round until, all at once, the illusion of contentment which the day had fostered was suddenly dispelled by a flash of memory. I saw again the steady, deadly, hypnotic pacing of a timber wolf in a cage . . . hour after endless hour . . . the pointless and repetitive circling of a prisoner.

In mid-morning the little police launch cautiously nosed her way into the Pond. She anchored and Onie put me aboard for a word with Danny and the constable. As we sat on the warm deck watching the whale, Danny told me that a considerable commotion was developing in Burgeo.

"Them radio broadcasts sure stirred 'em up! They's a bunch would like to run you over the end of the government wharf. Mad as hell at you for sticking your nose in where it waren't wanted. But they's plenty others thinks you done right. Trouble is, most of they aren't talking much, and the first lot talks *too* much."

"What do you think, Danny?"

Danny's sardonic face eased into a slight grin. "Well now, you're a bloody fool, a-course. Still and all, 'tis toime that trigger-happy crowd was slapped in the chops. Getting worser every year, they is. Was a toime, not long since, when a man went gunning if he needed meat. Now, be Jasus, I t'inks some of them carries a gun to the shithouse . . . in case they get a chance to shoot the neighbour's cat!"

He paused and eyed the circling whale reflectively.

" 'Tis a quare thing. I hears the Sou'westers Club is all for the whale now. Going to feed it and cosset it. But I guess I knows why. Might be good for business. Might help get the government off its rump and put a highroad into Burgeo. But here's the joke, Farley, bye . . . a good part of that lot was right here pumping lead into the poor jeezly whale less'n a week ago."

The launch departed and once again we three were alone with the whale, though not for long. An Otter seaplane roared high overhead and landed in the direction of Short Reach. Soon afterwards a power skiff entered the Pond and unloaded a C.B.C. television crew which had been flown in from St. John's. I knew most of the crew and was pleased as well as surprised to see them.

"Where's your cockeyed whale, Mowat?" the lanky cameraman asked as he struggled up the ridge toward me. "Or did it come out of the bottle too many you drank last night? My God, you've started something! Columbia Broadcasting System's flying a crew up here from New York. Toronto put the bee on us to get them film for C.B.C.'s national newscast tonight, or else . . . so if you don't have a whale, you'd better get one goddamn fast!"

"There's your whale," I said, pointing to the Pond where she had just begun to surface.

The four men in city clothes turned to stare in the direction indicated. Somebody gasped audibly.

She had chosen to surface not more than thirty yards off the point where we were standing, and less than fifteen yards beyond Onie who was sitting in the drifting dory. As she floated up out of the deeps, her massive, green-tinged bulk seemed to be magnified to unbelievable proportions by the distortion of the water.

Current boils the size of swimming pools marked the thrust of her flukes. Then the sleek black dome of the breathing hump broke surface and a column of mist shot twenty feet into the air, hung like a diaphanous haze against the sun, and slowly began to fade as the entire length of the beast's back wheeled into view and sank below again. It was some time before anyone spoke; then the cameraman turned to me and his usually quizzical face seemed oddly solemn.

"Holy Mother of God!" he said softly. "You've got a whale!"

After that I was ignored. There was a wild scramble to set up the camera gear, and not until the last foot of film had been shot did anyone have further time for me.

While they were taking down the tripods, the producer offered me a drink from his flask. "You know, Farley, they're treating this whole thing as a kook story on the mainland. Ahab Mowat and Moby Dick sort of bit. Funny cartoons in the newspapers. Max Ferguson did a hilarious skit on national radio this morning about a fight between Prime Minister Pearson and Smallwood, whether it was a provincial whale or a federal whale. We thought it was a kook yarn too. Not anymore. Did you ever *see* anything so damned big? Poor bloody beast. I hope you save its life. I hope you get it out of here somehow."

One of the things I tried to do that day was to determine the damage done by the gunners. After the C.B.C. crew departed I joined Onie in the dory and we rowed some fifty yards off shore and then sat motionless. Because of the exceptional clarity of the water, it was possible to get a look at the whale's underbody when she passed close alongside. On one occasion she surfaced within a dory-length of us and I had the uncanny feeling that the gaze of her great eye—it seemed to be as large as a man's head—was directed straight at me. Certainly she must have had as good a look at us as we did at her. Time and again she passed directly beneath the dory, or a few feet to either side, as if she were deliberately courting our company.

The slice across the base of her fin, a handbreadth wide, showed a layer of yellowish-white blubber about six inches thick, with a dark red, almost black cut in the underlying muscle tissue. By comparison to this, her numerous other wounds seemed trivial. I counted about a hundred and fifty small white breaks in the black skin which were certainly bullet holes; but on that huge animal they seemed of little more significance than mosquito bites on a man. They were not bleeding and their apparent unimportance strengthened my hopes that the shooting had perhaps done her no serious harm. I was happy to believe that the vast bulk of muscle and bone beneath the blubber could doubtless absorb rifle bullets as easily as a bull might absorb a charge of shotgun pellets fired into his rump.

My optimistic assessment of her wounds, together with the evidence of her apparently undistressed behaviour, convinced me that the corner had been turned. With the outside help which I was now sure would soon be on its way, I believed we might win the battle for her life and freedom.

But it could not be won until we had solved the problem of feeding her. I could see for myself that she was rapidly losing weight. Her back was becoming V-shaped, and the bulges which marked the locations of her enormous vertebrae, increasingly distinct. Her rapid loss of weight, together with the absence of any very young whales in the family pod, strengthened my suspicion that she was pregnant. There was no way I could be sure about this, but I had to work on the assumption that she was carrying a calf, and one, moreover, that could not be much more than two months short of term.

The picture was not all dark. If herring were needed in quantity, at least they were to be found close at hand. The difficulty was to get them into Aldridges Pond and hold them there until they became whale dinners. One solution might have been to hire a few men to run gill nets in Short Reach and then haul dory loads of freshly netted herring into the Pond and dump them there. However, I had my doubts whether she would, or could, take dead food. A Fin Whale's head is formed for engulfing living, swimming schools of fish, or concentrations of plankton, in mid-water, and not for scooping dinner off the bottom, even assuming that a Finner would eat carrion at all.

The opportune arrival of the Hanns with a big load of cod from The Ha Ha gave me an idea. We rowed over to where they were cutting and I asked them to save the cods' belly contents, which consisted almost exclusively of herring. When they had finished, Onie and I took their big bait-box, which now contained about 200 pounds of dead herring, and, balancing it precariously across the forward gunwales of the dory, we rowed to a point off the islet where the whale was in the habit of surfacing.

We let her make several undisturbed circuits of the Pond and on the fourth, just as the green-white mass of her chin was rising out of the depths toward us, Onie pulled the dory into her path, while I tipped the bait-box overboard.

The herring sank belly up through the clear water, glittering like metallic confetti. The whale had only to open her mouth and accelerate a little in order to scoop in the entire mass. She did nothing of the kind. With an almost imperceptible motion of her flippers, she swung gently to one side, avoiding the cloud of dead food, and passing on to surface and blow a hundred yards away. It was an experiment which did not need to be repeated. Clearly she needed a supply of living herring which she could catch for herself.

Reluctantly leaving the Pond (for I knew the supperb weather could not last and I hated to waste any of it), Onie and I set off to visit the manager of the fish plant who at my request had arranged a conference with my new allies, the members of the Sou'westers Club.

I was a little nervous about my reception at the

plant which, according to Danny, had become hostile territory to whale-lovers. Leaving Onie with the dory, I walked to the office past a number of men whom I knew had been involved in the gunning of the whale. Although nothing was said, there was no masking their hostility. The atmosphere at the office was quite different. The manager greeted me warmly and called in several other senior employees who were also Sou'westers. We began discussing the problem of how to feed the whale. As a first step, the manager volunteered to have his men construct an open-work barge, ballasted so its hinged top would float just at water level, in which live herring could be transported. The plan was to tow the filled barge through the channel and release its contents in the Pond. The manager thought the barge could be ready the following morning.

This was fine, but we were still left with the basic problem of how to get the live herring with which to fill the barge. While we were mulling this over, one of the Sou'westers asked if I had happened to hear a radio address delivered by Premier Smallwood a couple of hours earlier, in which the Premier had said not only that the whale was now the official property of the province, but that the government intended to do everything necessary to save its life.

Because I had been at the Pond I missed this broadcast but, hearing it now, remembered a telegram received the previous day from a friendly sailor on the *Harmon II,* a herring seiner owned by the Newfoundland government and used for the training of outport crews. The *Harmon* was lying inactive at her berth in Cornerbrook on the west side of the island and the sailor wondered if she could be put to work catching herring for our whale. I told the Sou'westers about the telegram, and added:

"If Smallwood means what he says he can hardly refuse to send the *Harmon*. It shouldn't take her more than a day or two to get here. With her gear she can seine a hundred tons a day and we can turn Aldridges into the biggest damn herring bowl in history!"

"Might take some time to arrange, though," the plant manager said cautiously. "Government don't always move too quick, you know."

"Well, all right. In the meantime why don't we call British Columbia Packers at Hermitage? They're fishing at least a dozen seiners on the coast; some of them right here among the islands. We'll ask them to donate a few tons of live herring as a stop-gap until we get the *Harmon*."

I picked up the manager's phone and, after fuming through the usual delays, reached a B.C. Packers official in Hermitage. I told him our needs and waited expectantly while he took a few seconds to think it over.

"Sorry," he said at last, "we can't spare the herring. Need all we can get to keep our plant working full shift. Sorry about your whale . . ."

He hung up before I could become abusive. The plant manager calmed me down.

"Listen now. There just might be some old capelin seines around. We could find out easy enough. We could maybe hire a crowd and hand-seine into the cove at Aldridges tonight when the tide's high. Might be able to purse what herring there is in the cove and drag the seine right in through the gut. Then all we'd have to do was keep the entrance barred off till the whale had her feed."

Since this appeared to be the only course open for the moment, we decided to give it a try.

Chapter 15

I arrived home to be met by a distracted Claire. "Thank heavens you're back!" was her relieved greeting. "There . . . that damned telephone again! *You* answer it!"

The call was for Bob Brooks from his impatient editor, ordering him to depart from Burgeo that very night.

"Bloody fool," Brooks muttered after he rang off. "Does he think I'm going to call a cab, or maybe jump aboard the next Air Canada jet? Where in hell does he think I am?"

It was a good question. Most of our callers seemed to believe Burgeo was a suburb of Halifax, or maybe of Boston. The very impatient producer of a major U.S. network show had telephoned to inform Claire that he and his crew were catching the first scheduled flight to Burgeo and would be on hand next morning.

"Tell Mr. Mowat to have 110-volt power available beside the whale for lights, and we'll need two half-ton trucks and a station wagon to carry our gear from the airport."

Claire rose nobly to that one. In the most winsome accents, she replied: "We'll try to find you some gasoline lanterns. I'm sorry but there is no airline and no airport. If you can find a charter ski-plane to take you as far as Gull Pond we can probably get a dog-team to pick you up there. But do bring your snow-shoes, just in case."

As we sat down to a hurried meal, I leafed through the messages. There was another telegram from the Premier.

> I HAVE THE PLEASURE AND HONOUR TO INFORM YOU THAT YOU ARE APPOINTED KEEPER OF THE WHALE STOP THE OFFICIAL DOCUMENT RECITING YOUR APPOINTMENT WILL BE FORWARDED IN DUE COURSE STOP KINDEST REGARDS.
>
> J R SMALLWOOD

"What the devil is *that* all about?" I asked in bewilderment.

"Joey was interviewed about it on the C.B.C. this afternoon," Claire explained. "He said the whale is worth a hundred million dollars in free publicity to Newfoundland. Canadian Press wired a copy of an interview *they* did with him. Here, read it for yourself."

> St. John's, Nfld. C.P.
>
> Author Farley Mowat has been officially appointed Keeper of the Whale and an appropriate uniform for the office is under consideration, Premier Joseph Smallwood announced in the Legislature.
>
> "We have not decided upon a uniform," Mr. Smallwood said. "He normally wears a kilt. But I'm sure we would not want him monkeying around with an 80-ton whale wearing a kilt."
>
> As laughter rippled through the house, the Premier cautioned that no member or citizen should "take lightly the extension of this great tradition in Britain's oldest colony."
>
> He quoted The Keeper of the King's Purse and The Keeper of the King's Conscience as other examples of the office, adding, "It has now been extended to Newfoundland for the first time with this appointment."

"The whale has a name now, too," Claire said. "You'd never guess . . . it's Moby Joe! For good old Joey himself."

"It may strain his famous sense of humour a bit when he finds out his namesake is a lady, and probably pregnant to boot," I replied. "Well, let him have his fun. Main thing is he's officially said the whale is under government protection. I'm going to wire him to send the *Harmon* down."

Smallwood and the mass media were not the only ones to "climb aboard" the whale that day. There was a message from an entrepreneur in Montreal offering me $100,000.00 if I would deliver her, alive and in good condition, to the World's Fair, Expo '67. Another offer came from a circus owner in Louisiana stating he would be happy to buy the whale if I would have her stuffed or otherwise preserved for exhibition.

Canadian marine science now also awoke, if only partially, to its golden opportunity. A biologist, who was too busy to come and see the whale for himself, sent the following telegram instead.

> DESIRE OBSERVATIONS OF WHALE BY CRITICAL OBSERVER SUCH AS YOURSELF LIKE TIME IN SECONDS BETWEEN BLOWS OVER TWENTY-FOUR HOUR PERIOD STOP CORRELATION ACTIVITIES WITH A DECREASED BLOW RATE STOP FEEDING BEHAVIOUR IN DETAIL INCLUDING SPEED AND RADIUS OF TURNING STOP AVOIDANCE OF OBSTACLES AND DEFINITE PLAY BEHAVIOUR STOP HOW MANY EXCRETIONS OBSERVED TWENTY-FOUR HOURS . . .

Not all the responses from the outer world were as absurd. There were a number of telegrams from individuals whom I did not know, and would never meet, which were simple and moving affirmations of the fact that some human beings *could* care about the unhappy plight of a distressed member of another species. These messages were something of an antidote to the unpleasant suspicion that I had got myself involved in a public circus. And, too, they helped ease the guilt I was feeling at the brutal and unthinking assault upon the people of Burgeo as a whole by the media—an assault which I had been instrumental in unleashing.

Almost every story printed or broadcast during those first few days took pains to stress the attack made on the whale by the riflemen, making it sound as though the entire population of Burgeo had taken part in an orgy of bloodletting. Mnay writers and broadcasters took a tone of holier-than-thou revulsion against the barbarism of an uncouth band of savages.

For years I had been publicly extolling the virtues of people on the fringes of "civilization," whether they lived in arctic igloos, aboard ocean-going tugs, on prairie farms or on the coasts of Newfoundland. I had celebrated their unsophisticated honesty, defending them against the smug contempt of admass man. I had sought to be their champion—and now, because of my feeling of kinship for the whale, I was being made to look as if I had turned against the outport people and had joined their detractors.

The seining effort had been arranged by the Sou'westers during the afternoon and now they called me to say it would consist of the two Anderson brothers—a pair of dour little men who owned the only capelin seine in Burgeo; Kenneth and Douglas Hann; and Curt Bungay and Wash Pink. The Hanns and the Andersons would each provide a dory to assist in working the seine. Although high tide was not due until after midnight, I thought it would be wise to make a preliminary reconnaissance soon after dark. Curt agreed and we set off in his boat.

The sea, black and motionless as stretched silk, was literally alive with herring. Immense schools drove off on either bow. They were right at the surface and, as they dashed away from the boat, the ebony face of the waters suddenly glowed in pale bands of saffron phosphorescence. Occasionally one shoal overrode another and thousands of herring broke through the surface and sparkled in the beams of our running lights like myriad shards of mirror glass. When I turned the spotlight downward, it revealed layer upon layer of silvered fishes as far as the light could penetrate.

Only once before had I witnessed such a stunning aggregation of living things. That was in 1947 when I watched the mass migration of tens of thousands of Barrenland caribou over the Keewatin tundra. At the time it had seemed inconceivable to me that anything could diminish such multitudes of living creatures; yet, before a decade had passed, the caribou had been virtually eliminated from much of their immense arctic range.

On that winter night in 1967, it seemed inconceivable that such vast numbers of herring could ever be significantly diminished even by man's monstrous predation; and yet by 1972 those vast schools had been so heavily decimated that informed biologists were predicting an end to the herring fishery in the entire North Atlantic before the end of the decade.

The little cove outside the Pond was jammed with herring. "Lard Jasus!" cried Curt with the uninhibited enthusiasm of a true fisherman. "Would you look at that! I don't say as what a man couldn't git out and walk ashore and never wet his feet!"

For fear of disturbing the schools, we did not enter the cove but instead landed a few yards farther down the open shore. With the aid of flashlights we scrambled over the slippery rocks to the channel in order to remove the barrier net the Hanns had put in place. Someone had been before us. The head-rope of the net had been cleanly sliced and the net itself had already been hauled ashore in a tangled heap.

When I asked the Hanns for an explanation of the ruined barrier net, they were evasive. Not until later, when Curt and I were alone in his kitchen awaiting the arrival of the Andersons with the seine, did I get an explanation.

"Seems as if some of the people down to The Reach is right ugly about you barring off the Pond," Curt said. "Claim nobody got the right to bar off a boat passage. Not you, nor the Mountie, nor Joey hisself."

"But," I protested, "the Hanns fixed the net so anybody could slip one end of it and pass a boat through with no trouble. The fishermen must know it's not intended to keep them out of the Pond. It's just to keep the herring in."

"No matter, Farley. 'Tis partly that sign you put up saying the Pond is closed. We . . . them fellows been free to come and go on the water anywhere they wants, all their lives. If you was to bar the channel with anchor chain I don't say they wouldn't cut it clear somehow."

I was getting angry.

"That's just damn stupid! Surely they realize we have to hold herring in the Pond. And it won't be forever . . . not more than a month at most . . . Well, the hell with it! The channel's going to be barred and it's going to stay barred so long as need be."

Curt's round, red face was impassive and he made no reply. I got to my feet. "Let's get the crowd and go. It's midnight and we've work to do."

The moon was up by the time our three-boat flotilla deployed off the mouth of the cove. Quietly the men fed the hundred-yard-long seine over the stern of Curt's boat as she slowly moved parallel to the shore. The Andersons, in one dory, were left to hold the free end. When the net was all paid out, the second dory, manned by the Hanns, took the remaining end. Then both dories moved landward toward the cove.

The curved net began inching shoreward, as Curt, Wash and I waited, watching silently. Then, as if there had been an explosion down under, there came a violent eruption and a surge of water that set us rolling wildly. I instinctively grabbed for the gunwales to brace myself just as a sonorous *whoosh* sounded so close at hand I thought for a moment it was in the boat itself. A fine mist settled over us. Transfixed, we stared astern where the streaming back of the Guardian was arching smoothly under, not a dory's length away. Curt was the first to recover.

"Lard God Almighty! That was close enough!

Hope the buggar's got his radar working! What the devil do you suppose . . . ?"

He was interrupted as the usually imperturbable Wash Pink shouted:

"Look, bye, look there! *Look* at them herring drive!"

Fifty feet to port, between us and the entrance to the cove, the water was surging with the bodies of wildly fleeing herring. In the glare of the spotlight it looked like a groundswell running to shore after a storm. The swell seemed to break against the right-hand section of the seine and along the remaining gap between the Andersons' dory and the shore.

Curt and I looked at each other.

"You don't think . . . ?" I began hesitantly. "You don't think he did that on *purpose,* do you?"

"Purpose or no, Farley, bye, he sure and hell druv a couple hundred barrels right straight into the cove."

Later the Hanns told us the rush and flow of herring under and around them had rocked their dory. Kenneth had no doubt at all as to what had happened.

"That whale you calls the Guardian, we t'inks he has his own way of doing for the one inside. Me and Doug've seen he rush the cove like that afore. 'Tis certain he be trying to drive herring into the Pond for she to eat."

Fortunately for our nerves, the Guardian did not reappear. In a few minutes the two ends of the seine were in shoal water. Wearing hip waders, the four fishermen went over the sides of their dories and began pursing the seine through the cove toward the mouth of the channel. Penned ahead of them was what we later estimated to have been about five tons of milling little fishes.

When the trapped school was almost in the mouth of the channel, the men began wading back and forth along the perimeter of the seine, banging the surface of the water, shouting, and shining flashlights into the seething mass. Finally an arrow stream of fishes sped into the channel and in a moment the entire school was pouring into Aldridges.

It was too dark to see what was happening beyond the channel, but the swoosh and surge of water, and a mighty splashing of flukes, told us that the prisoner was hungrily enjoying the bounty we had provided.

The first sweep was a great success, but a second failed when the seine snagged on the rocky bottom and allowed most of the herring to escape. Nevertheless, the men managed to bag four or five barrels in one end of the net, and they decided to tow the bag right through the channel. Kenneth Hann described what followed:

"We was no more'n ten feet inside the Pond when we opened up the bag and let the herring go free. I had a holt of the seine and was leaning overside to see was they all clear, when Dougie yelled for me to look arter myself. I turned me head and there was the whale coming straight for we with her mouth open wider'n the main hatch of the *Baccalieu*. She was pushing water ahead of her like one of them ocean liners, and the dory rose right up and near capsized.

"By the time I got me bearings she was gone again. And I tells you, bye, it never took we long to be gone out of there ourselves. If they was a herring she never got, he must have been some slick. She could have had we just as easy, but she never bothered. 'Twas just as well. I never did envy that fellow Jonah anyhow."

All in all it had been a good day. Temporarily, at least, we had solved the feeding problem. And we knew for sure that the lady in the Pond had a good appetite and was able and anxious to fill her belly.

Chapter 16

I woke late, to the familiar whine of a sou'west wind and the dry rustle of snow blasting against the house. Already, great grey seas were pounding the headlands of the offer islands. It was clear there would be small chance to visit the whale this day. I viewed the storm with mixed feelings. We could do no further seining while the sou'wester lasted but, on the other hand, the whale would be left in peace, free of human interference and intrusions, to feed on the herring we had already driven into the Pond.

But if she was to have a day of peace, we were not. Before I had been up ten minutes I was called to the phone, and for the balance of the day was engaged in dealing with an almost continuous telephonic siege.

Many of the calls were querulous requests for help from newsmen and camera crews stranded in storm-bound airports all the way from Halifax, Nova Scotia, to St. John's, Newfoundland. Some of them seemed to take the storm, and Burgeo's isolation from the world, as personal affronts. How was it, asked one television interviewer, with an edge in his voice, that he had recently been able to cover stories in Helsinki, Tokyo and Chile, all within the space of a week, but couldn't reach a godforsaken little burg in his own country?

A bit impatiently I explained that he could always do as we natives did and catch the weekly steamer to

Burgeo from Port Aux Basques. This suggestion seemed to give him little comfort.

It also gave small comfort to Bob Brooks. He was now in a fever of impatience to be gone with what was still a photographic scoop on the whale. However, although the westbound coastal boat was due in Burgeo this day, he could not be persuaded to book passage in her. Apparently modern media man, deprived of winged machines, is something of a cripple. And it was obvious that the only wings we were going to see over Burgeo for the next day or two would belong to storm-tossed gulls.

I did not feel much sympathy for the problems of the news people but it was quite another matter when I got a call from Dr. William E. Schevill at the marine research station, Woods Hole, Massachusetts. I knew Schevill to be one of the world's foremost whale biologists. He told me he had heard about our whale and was ready to come to Burgeo at once. His difficulty was how to get to us. I suggested he try to arrange something with the U.S. Naval air base at Argentia in southeast Newfoundland. This he did with such dispatch that he was able to call me from Argentia late that evening with the news that not only had he reached Newfoundland, but he was standing by with an amphibious Navy aircraft at his disposal, ready to descend on Burgeo the moment the weather cleared.

"It may not clear for days. If you want to be sure of getting here, you'd better come by sea," I told him, adding half-facetiously that he might ask the Navy to lend him a destroyer.

Alas, the Navy was not to be persuaded. The storm had worsened and the Naval commander at Argentia decided that the roaring rock wall of the Sou'-West Coast was no place for one of his ships.

The Burgeo whales seem to have decided that the inshore waters were no place for them that day either.

Early in the morning the Canadian government ship *Montgomery*, which maintained and serviced lighthouses and navigation aids on the coast, put into Burgeo for shelter. She dropped two anchors in the mouth of Short Reach only a few hundred yards from the entrance to Aldridges Pond and there she rode out the gale for the next two days. One of her people, an oceanographer, made the most of this opportunity to watch the Burgeo whales.

Shortly after the *Montgomery* came to anchor, a pod of four appeared in the mouth of The Reach. They were fishing hard, as if perhaps they knew it might be some time before they would feed again. Time after time they worked shoals of herring in toward Greenhill Island and the oceanographer could see sudden flurries of little fishes break the surface at the end of each whale's rush.

About noon, by which time the sea was building to formidable size, three members of the pod swam past the ship in close formation, heading resolutely southeast toward open water. When their spouts were last seen they were nearly a mile offshore.

"I guessed what they were up to," the oceanographer told me later. "We'd heard the marine forecasts and we knew there was a devil of a blow coming. Our skipper had to choose between running in to the lee of the land for shelter—which he did—or heading out to sea to get as much of an offing as we could. And I think the whales must have had just about as good an idea of what was brewing as we did. Only *their* choice was to go offshore. It made sense. Even in as good an anchorage as ours, the groundswell was bad enough to keep the ship tugging at her anchors like a wild horse. The surge coming right up from the bottom would have made lying uncomfortable for a whale. But once away from land they could go way below the turbulence and drift around in comfort with just the occasional trip back up to spout. It's the same principle submariners use in a storm: go deep and find the quiet water."

Though three of the pod made this choice, one of the whales remained "on station," as the watcher put it, seldom more than a quarter of a mile from the mouth of Aldridges Pond.

"That would have been your Guardian whale, I expect. And I'll tell you, he never shirked his duty. Except when the snow drift got too thick, or at night every time I looked that way I'd soon see his spout. It was curious how often he spouted, too. When the pod was fishing together they only blew about once every ten minutes, but this chap was up and down every two or three minutes and he'd sometimes stay up for quite a while, usually when he was right at the entrance to Aldridges Pond. I got the idea he was thinking of having a go at the channel himself, though I never actually saw him try it."

While the rest of the family went seaward, leaving the Guardian at his post, the lady whale remained unmolested in the Pond. But when Danny and Murdoch made a patrol in mid-morning, before the weather got too rough, they found that the barrier net, which we had replaced the previous evening, had again been cut adrift and badly mutilated.

"The whale seemed right contented," Danny reported. "We stayed about half an hour and she swam right up to we, as if she recognized us and was glad to have a visit. Once or twice I thought she was going to rub along our side like a cat'll rub against a man's leg. Thank God she didn't do it. She'd have heeled the old launch right over on her side."

Danny had one disquieting thing to say. "Remember when we first saw her, her back was smooth as a baby's bum? Now she looks sort of lumpy over pretty near the whole of her topsides. Don't know what to make of it."

Without giving it much thought, I assumed this "lumpy" appearance to have been due to a continuing loss of weight, and it made me even more concerned about the food supply. There was no way of telling when our barrier net had been cut, or how much of the herring school had escaped.

Since by Thursday noon Premier Smallwood had

not replied to my request for the *Harmon,* I sent another even more urgent wire in which I also asked that the R.C.M.P. be empowered to protect the barrier net and to restrict boat movement in the Pond.

Although there was no direct response from the Premier, rumours began to filter in from St. John's to the effect that the *Harmon* would be sent; that she was already on her way, and even that Smallwood himself would make an official visit to his ward on Saturday.

As the gale increased during Thursday night, so did my problems. Curt Bungay stumbled through the blizzard, his crimson face positively blazing from the wind, to tell me that the Hanns and Andersons wanted to be paid immediately for their work.

"Wash and me," Curt explained in some embarrassment, "we's satisfied to wait, but the others says on account of Joey give you a t'ousand dollars for whale feed, some of it rightly ought to go to all of we right now."

"But I haven't got the money yet. It's only promised."

"I believes it, but they fellows thinks you got it in your pocket. They won't believe no different."

"How much are they asking?"

"Well . . . they figures each man should get $25.00 for each fishing, and $10.00 for each dory, and $20.00 for the seine."

"That's $200.00 a time!" I protested.

Curt nodded. " 'Tis high, that's certain . . . but they's lots of hard feelings in Burgeo right now. Some says it's a cruel waste for the government to give money for to feed a whale at all. Some says . . ." he paused, reluctant to continue.

"What do 'some' say?" I asked grimly.

"Well, Farley, bye, 'tis this way. Some says you and that Sou'wester crowd is only looking out to yourselves. They can't stomach the way some of they fellows shot at the whale one day and set theyselves up to save it the next. They thinks you fellows is going to pocket whatever money comes."

"Anything else?" I snapped.

Curt was almost too upset to answer. However, he also had a temper, and it was rising.

"Well . . . yiss . . . since you asks. There's people so ugly at the closing off of the Pond, they says they'll finish off the whale afore they lets you bar they out. They got their rights, you knows. Nobody takes easy to it when they gets their rights took from them!"

The defiant note in his voice gave me pause. "All right, Curt, I'm sorry. Nobody's going to stamp on anybody's rights. You can tell the crowd I'll pay them what they want only they'll just have to wait until the money comes."

"Me and Wash'll wait, sure, and no charge for our boat neither . . . but I don't know about them fellows." Curt stood up and began buttoning his pea jacket. Then, hesitantly, he fished out a damp piece of paper and laid it on the table. "Don't know as you see this . . . 'twas posted up at the plant today."

With a quick "goodnight," he was gone into the storm, leaving me to study a mimeographed hand-out.

Citizens of Burgeo

Dear Neighbours:

You are well aware that in the past week or so there has been more publicity about Burgeo that there has ever been before. At this time when the citizens of the town and their elected representatives are doing everything in their power to get urgently needed facilities—i.e., Water, Sewer, Streets and Highways—this publicity is of immeasurable value.

Moby Joe is and will continue to be one of the most important "inhabitants" of our town as long as we can maintain it. As more and more people learn that Burgeo has a whale, we will have more and more people coming here to see and study it. The more people who come, the more important the town is and the more imperative it is that those in authority provide the facilities which are so urgently needed.

Will you cooperate with your neighbours in

their efforts to keep Moby Joe alive and healthy? You can do this by not using your boat in the Pond for pleasure cruising. We do not ask fishermen or those carrying water to stay out because they usually do not disturb the whale. It is imperative though that speedboats do not enter the Pond since this does scare the whale and could cause him to ground.

We, your neighbours, are depending on you, and we trust you will cooperate.

By all means go and see your whale. If you go to Richards Hole and cross over from there, you will have a perfect view and you will not disturb Moby Joe.

Sincerely yours,
THE SOU'WESTERS

I was annoyed, for I could see how this flier, a copy of which was later mailed to every family in Burgeo, would fan the hostility against the Sou'westers, with whom I now found myself bedded.

My mood was lightened by a phone call from Marie Penny, the "Queen of Ramea," as she was affectionately called, a widow woman of formidable ability and almost equally formidable presence, who owned and operated a small fish plant on Ramea Island, some fifteen miles from Burgeo. Marie was an old friend of ours.

"Hear you've got a new pet, Farley? Having trouble feeding it, are you? Asked Joey for a seiner, eh? Heard it on the radio just now. Well, boy, my guess is your whale'll be dead of old age before you see the *Harmon*. You should know better than to trust a politician! Now then, we've got a big capelin trap over here. Cost us $5000.00 and it's as good as new. Should work as well on herring as on capelin. I'll put it aboard the *Pennyluck* when she gets back from Hermitage where she's holed up waiting out the storm. You get somebody with some sense to set it and your whale will have a stomach ache from overeating. No . . . no . . . never mind the thanks. Just see you don't tear the bottom out of it."

Marie's was the last call that night. The telephone
went dead and we were able to go to bed. But the con-
flicts in my mind kept me from sleeping. I lay listening
to the thunder of the seas and the pulsating howl of
wind, feeling the house quiver on its granite rock.

What, I wondered, was the whale doing in the
bitter darkness of this raging night? What had she felt
during the long days of her captivity?

Pain, she had surely felt—and fear. Had she felt
despair? Did she have any hope of eventual escape? As
she circled the confines of the prison Pond, did she
ponder the horror of her probable fate? What wordless
thoughts were passing between her and the Guardian?
What did she feel about the two-legged beasts who at
first tried to kill her and who were now driving herring
into her prison?

No answers . . . none. Her mind was as alien to
mine as mine to hers. Strangers . . . strangers . . . we
were *all* aliens, one to the other, even those of us who
were cloaked in the same fleshly shapes. What did
I really know of the innermost feelings even of my
Burgeo neighbours . . . or they of mine? What did the
world beyond Burgeo know, or care, of the passions
the whale's coming was unleashing in this community?
Was there any real comprehension or true communication even between the human actors involved in this
bizarre drama? The more I thought about it all, the
more I realized that the inter-human conflict would
grow worse for want of understanding. It might well
become intolerable. Suddenly I wanted nothing quite
so much as to see the trapped whale freed . . . not only
for her sake now, but for my own ease as well. I
wanted her away from Burgeo where her presence had
become a threatening shadow of disruption.

. . . I dozed, and dreamed vividly of the whale.
She had become a veritable monster and I was flee-
ing from her . . . drowning in the unfamiliar element.
I woke, sweating, and knew the truth.

The whale was not alone in being trapped. We
were all trapped with her. If the natural patterns of her
life had been disrupted, then so had ours. An awesome

mystery had intruded into the closely circumscribed order of our lives; one that we terrestrial bipeds could not fathom, and one, therefore, that we would react against with instinctive fear, violence, and hatred. This riddle from the deeps was the measure of humanity's unquenchable ignorance of life. This impenetrable secret, which had become the core of our existence in this place, was a mirror in which we saw our own distempered faces . . . and they were ugly.

Chapter 17

On Friday the gale built to hurricane strength. By evening the anemometer was registering gusts of 80 miles an hour. During most of the day it was hardly possible to leave the house, let alone visit the whale. We remained apart: she in her prison, I in mine.

I spent the morning considering ways to free her. At noon I telephoned some friends in the Canadian Navy at Halifax. Between us we worked out a scheme. If I could get clearance from Defence Headquarters in Ottawa, they told me, the Navy would be happy to send in a team of frogmen to manhandle the boulders out of the south channel and deepen it as much as possible. (We could not risk using explosives.) With the next spring tide, which would come in three weeks' time, a section of steel-mesh anti-submarine net, floated by 45-gallon drums, would be rigged across the Pond like a gigantic seine to shepherd the whale toward the entrance. If she refused to attempt the passage on her own, I was prepared to use tranquillizer darts which would immobilize her so she could literally be hauled, or pushed, through the channel by main force.

I knew that the use of drugs would be terribly risky because too large a dose would render her unconscious and she might drown. So I phoned a whale expert in California who had had experience tranquillizing large porpoises. He was appalled by the magnitude of the undertaking.

"It'll have to be by guess and by God. I can tell you what specific drugs might work and help you get a supply by air, but I can only guess at the dosage. It's the hell of a chance, but if it's that or letting her die where she is, I guess it's worth trying."

Finally, I again called Jack McClelland in Toronto, explained what I had in mind, and asked him to obtain the cooperation of the Department of Defence. He groaned, but agreed to do his best.

I also called an officer of the Sou'westers Club to ask if they and the town council would put additional pressure on the Newfoundland government to send the *Harmon*. I had heard a radio report, which later proved false, that she had been dispatched to Burgeo but had been recalled because of the storm. Since the forecast called for a fine day on Saturday, I could see no reason why she could not reach us by Saturday evening. The Sou'westers' spokesman agreed to speak to the council but when I went on to describe the plans for freeing the whale, he became noticeably cool.

"Why turn her loose?" he asked. "She's doing good in the Pond. When we get the *Harmon* we can put enough herring into it to feed her for months. We ought to keep her there. No place else in the world's got a tame Fin Whale. It'd be a sin to turn her loose when she can do so much for Burgeo."

I was learning to be cautious, so I only mumbled something noncommittal and rang off. There was no point in my adding more fuel to the fires that were already smouldering in Burgeo.

I tried to shut my mind to what was happening in the community and concentrate on the three essentials: keeping the whale fed; keeping her protected; and arranging to set her free when the moment came.

By midnight Friday there was still no message from Premier Smallwood about the *Harmon* and I was beginning to despair of ever seeing her. However, I hoped we could repeat the seining operation at the cove on Saturday night, and by Sunday I expected to have Marie Penny's capelin trap in operation.

As for protecting the whale . . . in the continued

absence of any firm orders to the R.C.M.P. constable, this would have to remain my personal responsibility and I decided that the only way I could ensure her safety was to camp out at the Pond. The Sou'westers agreed to erect a small tent for me on the shore at Aldridges, and supply it with a stove and fuel. It could also serve as a field headquarters for the scientists if, and when, any of them should arrive. Schevill was still trying hard to reach us. He called at one o'clock in the morning to say the U.S. Navy was now planning to fly him to Burgeo before noon on Saturday.

In the outer world, interest in the whale had flared far beyond anything I had anticipated, and was becoming a nuisance. Radio stations from as far away as Texas and Colorado tried to tape interviews with me on the crackling telephone. There were wires from a Swiss newspaper syndicate and a magazine in Australia demanding information.

During Friday night the storm blew itself out, leaving a silence that was almost palpable. Saturday dawned an arctic day, icy clear and fiercely cold, with near zero temperatures. It was ideal weather, at least in Burgeo, for the ski-and float-equipped planes that were poised in a great semi-circle to the west, north and east, ready to descend on us with their loads of scientists, media people and, possibly, even Joey Smallwood himself.

One of the aircraft standing by at Gander had been chartered to pick up Bob Brooks, whose editor had instructed him to get some aerial shots of the whale as he departed. I consented to this on the understanding that the plane would stay at least 2000 feet above the Pond, and would not make more than two or three passes.

Brooks agreed, and Onie and I dropped him off at the country path leading to Gull Pond before going on to Aldridges. We had to sheer through a skim of cat ice that was forming in the protected runs between the islands, and the possibility that the Pond itself might freeze occurred to me. It was not an immediate danger since I expected that an eighty-ton whale would be able to break its way through a considerable thickness of salt-water ice.

Near the mouth of Short Reach we saw the spouts of at least two whales and assumed they were members of the family pod returned from their deep-sea sanctuary. As we had come to expect, the Guardian was in his usual place, patrolling between Fish Rock and the cove. We had grown so used to his presence—and perhaps he had grown so used to ours—that when our courses threatened to meet, Onie did not even slow the engine, but the whale sounded in good time to avoid a collision.

Onie smiled his gentle smile.

"That one knows the rules o' the road, Skipper. Vessel on t'port tack always gives way."

"Maybe he does, Onie, but I wouldn't push it too far. I'd hate like hell to have to swim ashore today."

There were no people at the Pond when we entered. The whale was circling in a counter-clockwise pattern and it was soon clear that something was amiss. Her movements were sluggish, lacking the powerful and fluid grace of earlier days. Also, she was blowing at very short intervals and her spout seemed low and weak. When she lethargically curved her way past our perch on a shore-side cliff, we saw that the full length of her spine showed clearly in a chain of knobby vertebral projections.

Another thing that troubled me was the presence of an irregular pattern of great swellings that showed under her gleaming black skin. Onie thought these might have resulted from bruising contacts with underwater rocks when she was being pursued by the speedboats or when she was trying to escape through the channel. I doubted it, but could think of no better explanation. Those peculiar swellings worried me, so we got back into the dory and rowed out to take a closer look. We let the dory drift directly in her path, for we had no fear that she would strike us, either deliberately or by accident.

On her first circuit she changed course slightly and passed fifty yards away but on her next she came straight for us. When she was about a hundred feet off, she did something we had seen her do only a few times before and then always at a distance. She rose to blow, but instead of breaking the surface with her hump, she thrust her whole head high out of the calm waters. The gleaming white, deep-pleated expanse of her throat, with its curving and apparently endless jaw line, seemed to belong to a creature three times her actual bulk, for such are the proportions of a Finner's head to the rest of its body. That gigantic head appeared to rear directly over us, like a moving, living cliff.

It might have been a moment of terror, but it was not. I felt no fear even when her eyes came out of water and she swung her head slightly so that one cyclopean orb looked directly at us. She had emerged from her own element as far as she could in order to see us in ours, and although her purpose was inscrutable, I somehow knew it was not inimical.

Then she sank forward and her head went under. The hump appeared, she blew and sounded and, a few seconds later, was passing directly under the dory. It seemed to take as long for the interminable sweep of her body to slip by as it does for a train to pass a railway crossing. But so smoothly and gently did she pass that we felt no motion except when the vast flukes went under us and the dory bobbed a little.

It was then I heard the voice of the Fin Whale for the third time. It was a long, low, sonorous moan with unearthly overtones in a higher pitch. It was unbelievably weird and bore no affinity with any sound I have heard from any other living thing. It was a voice not of the world we know.

When the whale had passed on, Onie sat as if paralyzed. Slowly he relaxed. He turned and looked at me with an anxious and questioning expression.

"That whale . . . she spoke to we! I t'inks she *spoke* to we!"

I nodded in agreement, for I will always believe she deliberately tried to span the chasm between our species—between our distant worlds. She failed, yet it was not total failure. So long as I live I shall hear the echoes of that haunting cry. And they will remind me that life itself—not *human* life—is the ultimate miracle upon this earth. I will hear those echoes even if the day should come when none of her nation is left alive in the desecrated seas, and the voices of the great whales have been silenced forever.

That Saturday morning at the Pond was a strange idyll. Nobody came near and nothing happened to break the companionable mood. Onie and I drifted in the dory, rapt, and unaware of the cold biting into our bones. The whale performed her slow and stately dance through the still waters, making a point of coming close to us as she completed each new circuit of the Pond. I cannot know how she felt, but *I* felt almost happy. I kept hoping to see her swirl into sudden action in pursuit of some small school of herring which might have entered the Pond, but it was a forlorn hope.

Just before noon I decided to go to the fish plant to see if there was any further news of the *Harmon,* or of the capelin trap from Ramea, and also to see that preparations for a further seining effort had been put in train. Onie had just opened the petcock, preparatory to starting the dory's engine, when we heard the nasal whine of an approaching aircraft. It was a ski-equipped Cessna which had slipped in to Gull Pond for Bob Brooks. Having picked him up, the pilot was now climbing over Aldridges where the plane began to describe wide circles as Brooks took his pictures.

I looked anxiously at the whale, but at first she seemed unperturbed. The plane made five or six circles at a reasonably high altitude, but then, instead of going away, it began to descend, circling lower and lower until it was snarling across the Pond at less than three hundred feet. I sprang to my feet and shook my fist at it, screaming useless imprecations. The plane ignored me, making pass after pass over the Pond while the engine's roar reverberated deafeningly between the rock walls of the hills.

The whale became panic-stricken. At what must

have been almost her maximum speed she burst down the centre of the Pond toward the channel, turning only at the last possible moment, in a wall of foam, before racing back toward the shallow eastern arm.

Onie was also on his feet and he, too, was yelling which was an act of which I had scarcely thought him capable.

"She be going to ground! Lard God, she be going to run ashore!" he cried in a voice sharp with fear.

Had I possessed a rifle I think I would have tried to shoot down that plane. I was quite beside myself when, after forty minutes, Brooks evidently decided he had enough photographs and the plane climbed away and disappeared.

In retrospect I do not particularly blame Bob Brooks. I suppose he was only doing what any media man is paid to do: get the coverage and to hell with the local consequences. But I was still shaking with rage when we reached the fish plant and I entered the manager's office. He had no good news with which to allay my fury. He had tried to call the *Harmon* on the ship-to-shore radiophone but could not raise her. Finally he succeeded in raising another seiner in Bay of Islands and learned that the *Harmon* was still lying at her dock, apparently under no orders to sail anywhere.

There had been no word at all from Premier Smallwood. But Gander airport reported three ski-equipped planes laden with press and camera people standing by for take-off. The delay in their departure for Burgeo seemed to be due to uncertainty about the thickness of the ice on Gull Pond.

The only hopeful news was a message from Ramea saying that the capelin trap was on the dock there, and would be picked up and brought to Burgeo on Sunday by the refrigerator ship *Caribou Reefer*. However, that delay meant we would not be able to use it until Monday, and might not get a catch from it until Tuesday. I was afraid the whale could not wait that long. The Sou'westers Club had agreed that we should mount another seining effort Saturday evening, but the Club officers were worried about paying for it.

I assured them that, if necessary, I would bear the cost myself.

When Onie and I returned to Aldridges we found the solitude had been broken. There were ten or fifteen boat loads of curious people present. Although most of them had left their boats in nearby Richards Hole, or in the entrance cove, one large trap skiff was dead in the middle of the channel, her propeller hopelessly fouled in the barrier net, which we had earlier replaced for the third time.

This boat belonged to a man named Rose who fished a little but whose preferred work was guiding mainland hunters who sometimes visited Burgeo to shoot moose. Rose had once come to me for help over a cancelled guide licence and I had been instrumental in getting it restored. I could hardly believe that he would now repay me like this. I shouted at him, asking if he had not read the sign asking boats to stay out of the Pond. Rose looked up at me, red-faced and hostile.

"I can't read!" he shouted back—and I was ashamed of myself for not rememebing that he, like many another outport man of his generation, had never had a chance to get much schooling.

"No matter *could* I read," he added defiantly, "I got me rights to go where I wants on the water, and 'tis what I aims to do!"

The timely arrival of the police launch saved the situation. Rose's boat was freed and he was persuaded by the constable not to motor into the centre of the Pond. But Murdoch could only use persuasion and, as the afternoon wore on, this was not enough. Several boat loads of young men challenged the police, and me, by roaring into the Pond. In our presence they did not quite have nerve enough to chase the whale, but they disturbed her sufficiently so that her erratic attempts to keep clear of them took her into dangerously shoal waters.

In contrast to the quiet of the early morning, it was now growing into a tense and miserable afternoon. Although most of the people present were seemingly content just to watch, the hostility between myself and the speedboat crowd hung over the Pond like a miasma. I was afraid that not even the presence of the policeman would be enough to restrain them, and I dreaded the possibility of another outburst such as had occurred on Sunday.

The tension was eased somewhat when a party of officers from the c.g.s. *Montgomery* came ashore to see the whale. The oceanographer was with them and he proved to be as sympathetic as I could have hoped. He told me the Guardian was still standing watch beyond the cove, but that he had sheered off and sounded when the press of visiting boats became too heavy.

Danny and Murdoch agreed to remain at the Pond until all the intruding boats had gone, since I had to return to Messers to organize the next seining expedition. I could only hope their mere presence would deter the "sportsmen" in the speedboats, and I was in a depressed state of mind as Onie piloted us homeward. He sensed it, and as he dropped me at Sim's stage, he said quietly:

"Don't you take it too hard now, Skipper. They's a good many people don't want that whale hurted. They thinks you's doing the right thing. Seems like Burgeo's gone adrift these times. People moved in from all along the coast, all mixed up together like mackerel in a puncheon. And some of they gone sour because of it, and don't rightly know what they's about."

The attempt at herring seining that evening was a fiasco. A porpoise expert from Florida had suggested that we try using lights to attract schools of small fish so they could more easily be seined. So I had borrowed a Delco generator from the fish plant and, when we reached Aldridges, I installed it on shore and set up a pair of lights which flooded the mouth of the gut.

The generator and lights worked well but when the men made their first sweep across the cove, the seine hooked on the bottom and tore to shreds. When I went aboard one of the dories to look at the remnants of it, I found I could pull it apart as easily as if it had been made of rotten straw. I could get no credible explanation from anyone as to why the net, which had functioned well enough only two days earlier, should suddenly be rotten now. When I asked Curt, he refused to answer or to meet my eyes. The Andersons, on the other hand, were defiantly defensive at the implication that they had substituted an old and useless net.

" 'Tain't the net's fault, bye. Nothing wrong with she. 'Tis hard use beat her up on the rocks. 'Tis only foolishness to fish a net on foul bottom. We only risked her to pleasure you, and now we's lost our net. Who's going to pay for it, we wants to know?"

Disgusted, I sent the Andersons on their way. The rest of us remained and kept the floodlights burning, hoping they would attract at least some herring into the Pond. A few barrels did in fact enter the mouth of the channel but they were reluctant to swim through it, and we could easily see why.

Undeterred by the brilliant glare, the lady whale had appeared at our end of the Pond where she swam back and forth as close to the inner opening of the gut as she could safely go. At times she came into such shoal water that she had to swim on the surface to avoid grounding. It was heartbreakingly obvious that she understood what we were trying to do and that she was desperately hungry.

It was two o'clock in the morning before we finally accepted the uselessness of our efforts and packed up the gear. I retained a faint hope that, in our absence, the Guardian might succeed where we had failed, and so we left the channel open.

It had been a long, distressing day. There had not even been the solace of seeing reinforcements arrive in the shape of Schevill and his party, for the u.s. Navy had decided that weather conditions were still not good enough to risk the flight. When I wearily stumbled up the steps and into the kitchen of our little house, it was to find Claire waiting with a table piled high with mail. During the afternoon the coastal steamer, s.s. *Baccalieu,* long delayed by weather, had finally reached Burgeo. Claire had walked to the post office and had staggered home under the weight of a full mail bag.

I was too tired to do more than glance at the mountain of letters from all over North America. Many of them contained small cheques and some held coins, gifts from all sorts of people: school children, the manager of a Chicago automotive works, a stockman from Calgary, a radio disc-jockey from New York, and a housewife from Labrador City. The gist of what they had to say was all much the same. They begged me, sometimes in extravagant words, to save and free the whale. Some were sentimental; but the words were of no importance. What mattered was that these scattered and diverse people in far distant places had all been moved by one thing, by compassion for a strange, great creature, trapped and endlessly circling in a small Pond on the remote coast of Newfoundland.

They gave me hope again.

Chapter 18

Schevill called early Sunday morning from Argentia to say that, despite the uncertain look of the weather, he and three other "whale watchers" would be taking off in an amphibious Catalina immediately. He wanted to know how landing conditions were in Burgeo.

"Good enough. Tell your pilot to find Short Reach on his chart. He can moor to the fish plant wharf."

"How's the whale?"

"Haven't seen her today, but we didn't get her a feed last night and she's getting desperately thin and awfully logy."

"Let's hope she'll make it. We'll be there soon to lend a hand."

The imminent arrival of professional support stirred me from a lethargy of fatigue and depression, and I bolted my breakfast, barely remembering to wish Claire "many happy returns" on this, her birthday. Soon there was a dull roar from eastward. We ran out on our porch and watched exultantly as a Catalina lumbered over Burgeo and began circling the village.

It circled and it circled.

"Why the hell don't they land?" I cried in a fever of impatience.

Almost as if he had heard me, the pilot put the plane's nose down and she began a ponderous descent toward Short Reach. As she slipped out of sight behind the intervening hills, I was already sprinting for

Onie's house to ask him to get the dory; but before I reached it there was a snarl of engines running up to full power, and the Catalina reappeared, climbing laboriously into the overcast.

I watched, incredulously, as she headed northwest toward the interior barrens and disappeared. When I realized she was not coming back, I went into the kitchen, glumly poured a cup of tea, and listened to the weather forecast on the marine radio. Strong southwest winds increasing to forty knots by midafternoon, with visibility lowering to zero in fog and snow flurries . . . I did not need the weather-man to tell me this was the beginning of another two- or three-day storm during which no aircraft would reach Burgeo.

Schevill called a couple of hours later from Stephenville, a "rented" U.S. Strategic Air Force base on the west coast of Newfoundland. It seemed that his Navy pilot had changed his mind about landing because he was afraid he might be marooned in Burgeo by bad weather. Rather than risk that awful fate, he had continued on to the comforts of the Stephenville base. Schevill was as disappointed as I was, but he remained optimistic.

"Never mind. I think I can talk the Air Force into bringing us in by helicopter. Look for us about noon."

We looked hard enough, God knows, but the helicopter never came. Instead, a Beaver on skis slipped in under the lowering overcast and landed on Gull Pond. It was some days later before we learned it had brought a television crew who, when they found they would have to hike cross-country from the landing Pond, got back into the Beaver and flew home again. No whale was worth *that* much effort.

By early afternoon we were experiencing a proper sou'wester and, since there was no longer any chance of visitors arriving, Onie and I set off through the grey storm scud for Aldridges Pond.

It was a forbidding sort of day. Streaming clouds, whipped by a whining wind off the black and frigid ocean, completely concealed the peak of Richards Head.

Snow flurries driving across the runs and tickles obscured even the familiar shapes of nearby rocks and islands. It was savagely cold, and the lop which was beginning to kick up in the mouth of Short Reach warned us we could not stay long at the Pond.

As we bucketed into the cove, visibility was so bad we did not realize we were not alone until we nearly rammed a whale, head-on. The whale was so deep inside the cove that there was hardly enough room for it to swim, let alone submerge. I glimpsed the gleaming mass of its head surging toward us when it was less than twenty feet away and, at my startled yell, Onie swung the tiller hard over and cut the engine. The whale also went into a hard turn, but in the opposite direction, and with such acceleration that the boil from its flukes heeled the dory far over on her side. Then another snow flurry swept down, obliterating everything from view. When the flurry passed, the whale had disappeared.

The wild thought flashed into my mind that, aided by a high tide raised higher still by the sou'-wester, the prisoner had escaped!

In jubilation, I yelled at Onie to start the engine. "I think that's her! I think she's *out,* Onie! Head into the Pond. Quick, man, quick!"

The old Atlantic barked into life and we shot through the narrow gut, propelled not so much by the engine as by wind and tide. The water in the channel was very high, higher than I had ever seen it before. Inside the Pond there was no sign of the trapped whale. I was now almost sure she had succeeded in escaping.

"Circle the Pond," I cried to Onie.

Obediently he put the tiller over and we puttered through the driving scud. As I stood in the bow peering about, I was vaguely aware that my initial surge of jubilation was fading and in its place was a growing and aching sense of anxiety; but I had no time to dwell on that. For then I saw her.

She was on the surface and moving very slowly. Almost all of her great length was exposed. She could easily have been mistaken for one of those colossal sea

monsters which decorate ancient charts. The illusion
was intensified by the vagueness given to her outlines
by the drifting snow.

I hardly know how to describe or explain my reaction. Instead of feeling sick with disappointment as
I realized she was still a prisoner, my spirits rose. I
felt something akin to elation. The only explanation
I can offer, and it is no easy one for me to accept, is
that if she *had* managed to escape without my help, it
would have made a travesty of my attempts to save
her. Or was it, perhaps, that I needed her continuing
presence in the Pond to justify my own actions and
attitudes toward those who had tormented her? Had I
come to rely on her presence in order to maintain my
rôle? Had *my* need of her become greater than *her*
need of me?

I have no answers to these questions, and I think
I want none.

She stayed on the surface an unnaturally long time.
Onie kept the dory running close alongside so we
would not lose sight of her in the snow flurries, and I
was horrified by the difference a single day had made
in her appearance. Not only had her back become
steeply and ominously V-shaped, but the inexplicable
bulges under her skin had grown much larger. There
was no longer an aura of almost supernatural vitality
about her—an aura which had strangely affected everyone who had seen her, including even those who
wished her dead. She seemed less like a living beast
than like some monstrous lump of flotsam.

She gave no indication of knowing we were so
close to her, but when she blew—a thin, almost instantly erased wisp of vapour—there was a sign, an omen.
The stench from her blow was a fetid assault upon
our nostrils.

At length she sounded, but slowly, as if with
great effort or reluctance. The snow scud streamed
down over the surface of the Pond, obliterating the last
faint swirl from her flukes.

Because of the increasing violence of the storm,

we could not stay with her any longer. Drawing our parka hoods close about our faces, we headed out through the channel into the cold fury of the storm. The Guardian, for it must have been he whom we had met in the entrance cove, was not in sight. As we bucked homeward through cascades of freezing spray, I thought about the encounter with him. I should have guessed that his presence so close to land, in such dangerously constricted waters, and on a lee shore, was also a portent. But I chose to believe he had simply been trying to drive herring into the Pond; and perhaps that *is* what he had been trying to do, although I now suspect he had an even more pressing urge to take the risks he did.

Claire and I had a little birthday party that stormy night, but our hearts weren't in it. A belated message had finally arrived from Premier Smallwood, informing me somewhat loftily that, although a Fin Whale could live six months on its stored blubber, he was nevertheless sending a certain Captain Hansen down by air to show us how to attract herring into the Pond with floodlights! Although he was not forthright enough to say so, it was obvious we would not get the *Harmon*.

God alone knew when, if ever, Schevill and his team of experts would reach us now. And there were rumours that the main herring run had already begun, prematurely, to move off the Sou'west Coast.

The cloud of foreboding I had been under during our visit to the Pond hung over Claire's birthday celebration which was, at best, a disjointed one. The telephone rang almost constantly as one unknown voice after another demanded fresh news of the Burgeo whale. When I could stand it no longer, I took the receiver off the hook and we went to bed.

When I awoke on Monday morning I was amazed to discover it was after ten o'clock. Sleepily I wondered how it was that the imperious demands of Mr. Bell's incubus had not dragged me from my bed at dawn. Then I remembered. Reluctantly I shuffled through the icy kitchen, automatically noting from the dial of the wind gauge that the gale had swung around into the nor'east. I restored the receiver to its cradle and barely had time to turn up the oil stove when the bell rang. A fisherman from Smalls Island, for whom I had once done a small favour, was on the other end of the line.

"Skipper Mowat? Is it you, bye? I'se been trying to raise you for a couple of hours now. We was out to The Ha Ha when the wind dropped out for a change at dawn, to see was our gear carried away by the starm, and the whale is beached. Aye, hard aground just inside the gut. 'Tis bleeding bad . . . looks to we like some'uns been at she with a lance. . . ."

Fear knifed through me more piercingly than the arctic cold brought by the nor'easter. I ran to the window and saw at a glance that no dory could live in such a sea. Frantically I called Curt Bungay and that good man agreed to chance the voyage to Aldridges in his decked motorboat.

The harbour waters were "feather white," but Curt was undismayed. He pushed his boat at full throttle until I thought he would drive her under. As he struggled with the wheel he yelled something into my ear that turned my fear to livid rage.

"Not surprised . . . was word some of they . . . going to strand she if they could . . . must have . . . lull this morning when . . . wind shifted round. . . ."

The fury that filled me verged on the homicidal. When Curt grounded the bow of the boat on the shore of the cove, I leapt to the slippery rocks so impetuously that I fell full-length into the landwash. Stumbling to my feet, I ran recklessly across the broken rocks of the intervening ridge. As I cleared the crest I saw her. She was lying directly below me. Her vast white chin was resting on the shore but, thank God, she was beached at a point where the water ran deep almost to the shore line so that most of her immense body was still afloat.

As I plunged down the slope toward her, I became aware of a foul stench—the same I had smelt the day before when she blew alongside Onie's dory. I also saw that the beach near where her head was resting was white with the partially digested bodies of herring. However, I saw these things without really seeing them, for I was totally engrossed in the urgency of getting her off that beach before the falling tide doomed her to die from her own great weight.

My memory of the next few minutes is hazy, but Curt, stumbling along behind me, saw it all and remembered the scene vividly.

"When I cleared the crest you was already on the beach. I could hear you yelling your head off before I even see you.

" *'Get off, you crazy bitch,'* you was yelling.

"Then the next thing I sees you was pounding on the head of her with your fists. You was acting like some fellow what's drunk too much white lightning, trying to launch a ship with his bare hands. Cause that's what she looked like to I. Like a ship with her bows ashore."

Curt was so dumbfounded and disconcerted by my behaviour that he remained on the ridge while I berated the whale and screamed imprecations at her, commanding her to shift herself. Finally, in utter desperation, I flung myself down with my back against an upright slab of granite, thrust my feet against the hard, rubbery curve of that immense mouth, and tried to shove her off by main force!

It was insane . . . a hundred and sixty pounds of puny human flesh pitting itself against the inertia of eighty tons of leviathan. Nevertheless, I pushed and I kicked and I yelled, and I may also have wept out of sheer frustration.

Then, almost imperceptibly, she began to move! I saw the flippers, big as dories, shimmer as they turned like hands on wrists. Slowly, so very slowly, she backed herself off the shore, turned, and cruised on the surface to mid-Pond.

Curt stumbled down the slope to join me.

"You done it, bye, you saved her, sure!" he cried. But I knew better. The scales were off my eyes, and now I saw the truth. She had not grounded by accident, neither had she been beached by the malice of men. She had *deliberately* gone ashore because she was too sick to keep herself afloat any longer. I had misread the evidence, but now it was unmistakable. There was the vomit stirring in the shallows where her head had lain. There was the stench. It was familiar now and well remembered . . . the same rancid stink which had made me retch away from the gangrene-rotted bodies of soldiers in Sicily in 1943.

There was more. As she moved slowly away from us she left thin ribbons of dark discoloration in the water. These were coming from the great swellings which had formed beneath her skin. I could see one of them pulsing out a dark flow of blood; and I realized that those swellings were vast reservoirs of pus and infection, some of which were breaking open to discharge their foul contents into the cold sea water.

As I watched, stunned and sickened, the whale continued to move across the Pond. She did not submerge. I doubt if she had sufficient strength to do so. Almost drifting, she reached the opposite shore and there she again rested her mighty head upon the rocks.

"Lard Jasus, she's beached again!" Curt shouted in alarm.

"No," I replied dully. "She's sick, Curt. She's too sick to even swim. If she stays in deep water she knows she'll sink; and then she'll drown."

Curt could not take that in. It seemed incredible to him that any beast which lived its life in the sea could drown. He shook his head in bewilderment.

The barriers of illusion had crumbled so suddenly that my mind was in chaos. How *could* I have been so blind as to believe she would suffer no real harm from the hundreds of bullets that had plowed into her flesh? Yet, in all honesty, how could I have imagined that this gargantuan creature might succumb to the attacks of infinitesimal microbes entering her wounds? Nevertheless, this was what was happening. What could I do about it now? Was it too late to do anything except curse my own stupidity?

I was in an almost paralytic state of indecision. I wanted desperately to get to a phone and talk to Schevill or anyone who had some knowledge of whale pathology; who could perhaps suggest what to do for the sick whale, and *how* to do it. On the other hand, although I now realized she had not been driven ashore by anyone, or lanced, I was very much afraid the news that she was beached and completely vulnerable would swiftly spread through Burgeo and convince some of her enemies that the time was ripe to finish her off. It was a measure of how deeply the virus of suspicion, anger and ill-will had entered into the human fabric of Burgeo, of which I was a part, that I was afraid to leave her unprotected.

That problem was solved by the arrival of the police boat after a hard punch out from The Reach. Danny Green had heard rumours that the whale had been attacked again, and he and Murdoch had risked the passage to the Pond. They anchored, and rowed ashore to join Curt and me upon the ridge. As I explained the situation, Murdoch stared across the leaden waters through his binoculars at the vast and motionless shape on the far shore. When I finished talking, he lowered the glasses and turned to me. His face showed how he felt . . . sickened at the sight of her; sickened and angry.

"We've still got no authority to stop people coming here," he said shortly. "But orders or no orders . . . no boat'll come near her again while we're about!"

I thanked him and turned to go . . . and then I heard the voice of the Fin Whale for the fourth time . . . and the last. It was the same muffled, disembodied and unearthly sound, seeming to come from an immense distance: out of the sea, out of the rocks around us, out of the air itself. It was a deep vibration, low-pitched and throbbing, moaning beneath the wail of the wind in the cliffs of Richards Head.

It was the most desolate cry that I have ever heard.

Chapter 19

On the way back to Messers we put in at Firbys Cove so I could collect a second bag of mail from the post office—mail from more well-wishers of the whale. As I hurried back to rejoin Onie at the dock, the heavy bag over my shoulder, I was confronted by a man I had known since my first arrival in Burgeo; a man for whom I had great respect and who, only a few days previously, had expressed his sympathy with the whale and with the attempts to save her. I greeted him warmly. He responded by deliberately spitting just to one side of my feet.

"What's that for, Matt?" I asked, bewildered.

" 'Tis for the likes of you! Strangers come here from away, telling lies about the people. Making troubles like we never had afore!"

He was a big man, and his words were delivered with such intensity I thought he was going to strike me. I stepped back; but he had no intention of using his fists. Words would serve.

"You and that bloody whale! Well, bye, she's finished now! And you're the same. Finished in Burgeo. I'll tell you that without a lie!" He turned on his heel and strode away.

Shaken by this outburst, I reached the dock where the dory lay and here was another unexpected confrontation. The two doctors were there talking to Onie. They looked up as I approached.

"Onie's told us the whale's sick," said the male doctor in a concerned and friendly manner. "Sounds as if it might be septicaemia. Is there anything we can do to help?"

I was astounded. From open advocacy of killing the whale, this couple had made as total a *volte face* as Matt . . . but in the other direction. It was all just too damned confusing. Those whom I had thought were my "natural" allies seemed to be turning violently against me, while those who were my "natural" antagonists were now offering to help . . . but at this juncture I would have accepted help from the devil himself.

"There might be something. What about antibiotic treatment for a whale? Is it possible? Could you give it?"

The wife, an aggressive and impetuous woman, answered.

"We could try. Only there aren't enough antibiotics in the hospital to make one dose for an animal that big. If you can get the drugs somewhere, we'll see if we can administer them."

I nodded gratefully. "Right. I'll see what I can do."

Twenty minutes after reaching home I had written a press release. It was an intemperate piece of work, reflecting the anger that I could no longer control. It began with a statement that the whale was probably dying of infection resulting from the wounds inflicted on her by the men of Burgeo. There followed the most harrowing paragraph I could compose describing the agonies the great animal was enduring—had been enduring for many days as her wounds turned septic. I concluded with a plea for help, for donations of antibiotics and injection equipment, "in order that we can try to make amends for the atrocious behavior of those who inflicted such tortures on the imprisoned whale."

I handed the paper to Claire to read before attempting to phone it out. She was horrified.

"You can't *send* this, Farley! It's . . . it's vindic-

tive. It's as vicious as what they did to the whale!
Please, don't send it." It was not sent.

When, about seven that evening, I finally got a
call through to Canadian Press, the story I gave was as
dispassionate as I could make it. C.P. promptly put it on
the Canadian wires and relayed it to the international
services. The C.P. district manager in Toronto took
the dictation personally, and when I finished he
thanked me and added:

"Moby Joe is front-page news across the conti-
nent. The story's stirred up the hell of a stink. It's
crazy, but people seem more worked up about your
whale than about the mess in Vietnam. I hope you
know what you're doing down there, Farley."

I was not sure what he meant by that parting shot,
but I did not ask. The truth was that I was no longer
sure what I was doing, or of what I had already done.
Fortunately, there was no time for reflection. Within
an hour the C.B.C. was broadcasting a special bulletin:

> Moby Joe's keeper tonight issued an urgent ap-
> peal for massive donations of antibiotics after it
> was discovered the trapped Burgeo whale had a
> huge infection in its back from bullet wounds.
> Farley Mowat said the whale was very sick. He
> said a husband-and-wife medical team in the
> hospital in Burgeo had volunteered to administer
> the antibiotics if they could be made available.
> They need 160 grams of petracyclin hydro-
> chloride for each dose and a minimum of eight
> doses will be required. They also need a three-
> pint syringe and a three-foot stainless steel nee-
> dle . . .

The response began to reach us almost immediately. A pharmaceutical manufacturer in Montreal
phoned to say that 800 grams of antibiotic was being
shipped to us from St. John's by charter flight at dawn
—weather permitting—and a further supply would be
flown from Montreal to Gander. A second message
told me that suitable syringes existed only at the Bronx
Zoo and at the Vancouver Aquarium, and that both

institutions had been asked to air express their syringes to Gander, from which point another charter flight would ferry them to Burgeo in the morning.

Schevill, still stranded at Stephenville, heard the first radio bulletin and spent hours on the long-distance phone consulting experts as far afield as Puerto Rico; obtaining opinions on the treatment the whale should receive, and setting into motion shipments of drugs from the United States. Then he called me.

"There's a good forecast for tomorrow. We'll make it in by helicopter in the morning, for certain this time!"

A veterinary surgeon from St. John's wired that he was flying to Burgeo at his own expense to give us a hand. Dozens, scores, of wires and phone calls plugged all South Coast circuits with offers of advice, encouragement and money. By midnight the response of the outer world had mounted to such a crescendo that the poor Hermitage operator, willing as she was, could not handle the flood. So we arranged to have a friend in St. John's accept and deal with the overflow.

The incredible and almost instantaneous response to the radio and television appeals had a curious effect upon me. The anger and grief of the early part of the day were submerged and washed away in an intoxication of excitement. The constant ringing of the telephone, and the blaring of radio voices describing the reaction to our plea for help, acted like a powerful stimulant. I felt like someone who discovers he can command miracles. I no longer doubted that I would save the whale. Realities were dimming in the euphoric glare of attention which played on Burgeo throughout that long, cold night.

Just before mignight I had a call from one of the Sou'westers. He was exultant.

"Are you listening to the radio? It's fantastic, eh? The old town's really on the map! Another couple of weeks like this and Joey'll be pushing the highway down to us. Thank the Lord for that whale! Moby Joe's going to put Burgeo into the modern times for sure!"

He paused, and when he continued there was a note of anxiety in his voice.

"She *is* going to pull through all right, isn't she?"

"She's sick and getting sicker," I replied. "Look, there's supposed to be at least five charter flights coming in early tomorrow with drugs and experts and I'll have to stick to the phone until I hear. Will you get someone to go to Aldridges as soon as it's light and keep and eye on things? The Mountie can't be there all the time, and I don't trust those bastards who peppered her. And ask someone to call me early on to let me know how she looks."

"Sure, Farley. Nothing easier. I'll go myself. Can't take a chance on something happening to her now."

It was another almost sleepless night for me. The angry tensions of the day, and the high excitement of the evening, had brought me to such a pitch that I could not even lie down for more than a few minutes. I kept the kitchen stove going and swilled endless cups of tea as I waited for the dawn.

When the first greenish tints of the new day washed the eastern islands, I went outside. There was hardly a breath of wind and, as the light strengthened, it revealed a cloudless sky. The fog-bank that always lurked a few miles off-shore, ready at any time to roll in over the land and smother Burgeo, was only an indistinct dark line on the far horizon. It was going to be a perfect flying day along the Sou'west Coast.

Claire was up and cooking breakfast when I went back inside. We ate almost in silence while she eyed me anxiously from time to time. Finally she said:

"Why don't you lie down for a little while. It'll be hours before any planes can get here and you've got to get some rest. It isn't a one-man job anymore, you know. There'll be all sorts of experts to look after her now."

The good breakfast, the red sun streaming low into the kitchen, and Claire's words, combined to ease my tensions so that I was hardly able to get up from the table. The worst was over. I felt as a man might who has stood a long and lonely siege and at last hears the distant sounds of an approaching relief column.

I lay down on the bed and instantly fell into dreamless sleep. Nevertheless, it was a light sleep and I was brought bolt upright by the harsh jangle of the telephone. Claire answered it. A moment later she was beside the bed.

"You'd better talk to them," she said in a stiff, almost frozen voice. "It's from the plant. The whale's gone. They can't find her anywhere."

It was exactly ten minutes past nine when I picked up the phone. I remember, because I automatically glanced at the kitchen clock to see how long it would be before the first plane could arrive.

"Farley? I'm just back from Aldridges. We was there just after dawn and we spent two hours looking all over the Pond and there isn't a sign of the whale. She's not there. She must have made a run for it last night. She's gone right out of it, boy. She's gone for certain."

"Gone?" I echoed stupidly.

And then I knew. I knew with absolute certainty.

"Gone? . . . She's not gone. She's dead."

My caller, who was a member of the town council as well as an officer of the Sou'westers Club, was not slow to grasp the implications of my flat statement. Probably he had already considered them during the search. Nevertheless, there was something close to panic in his voice.

"My God, man, she *can't* be dead! She *must* have swum clear! There'll be living hell to pay if the papers and radio get the idea she died here. They'll murder us!"

The whale is gone . . . the whale is dead. . . . The words echoed and re-echoed in my mind, and they lit the hard, white flame of hatred.

"You're right about that. Indeed you are. They'll murder you . . . just as Burgeo murdered the whale. Wouldn't you say that was fair enough?"

"Can't we agree to keep it quiet?" he pleaded. "If she *is* dead, she won't come to the surface for days in this cold weather. Can't we just say we think she's got free? By the time she floats, the whole thing'll have died down . . . You've *lived* here for five years . . . It's your town too!"

"No," I said. "It isn't my town anymore. I guess it never really was."

He was still expostulating when I hung up. I got Hermitage and, after the usual delays, the operator connected me with a newspaper reporter who had become the whale's unofficial agent "outside." I asked him to contact all those who were preparing to fly in to Burgeo, or who might already be airborne.

"Tell them it's all over," I said. "Tell them I'm sorry. Tell them they can all go home again."

"Farley . . . are you *sure* she's dead?"

"I'm sure. There's no way she could have escaped. She was so sick last night she could hardly stay afloat." And then, with an uncontrollable burst of bitterness, I lashed out at this good friend. "She's *dead,* do you hear me! Christ, do I have to rub your face in her stinking corpse to make you understand?"

He was a very good friend, and he forgave me.

Word spreads fast in the new world of technological wizardry. I had hardly finished putting on my parka when the program of music from the c.b.c. station to which we usually listened in the mornings was interrupted by an announcement that Moby Joe, the trapped Burgeo whale, had disappeared and was presumed dead.

Word spreads fast in the outports too. Even as the announcement was being made, the kitchen door opened and Onie came quietly in.

"I t'ought you might be needin' the dory," he said softly. "She be ready when you is."

The rambling, scattered and brightly painted houses of Burgeo; the wide-spreading, ice-encrusted islands; the glittering waters of the tickles and runs had never looked more beautiful than they did this morning as the dory made its way eastward. But now I was seeing it all as I had not seen it for many years . . . through the sudden eyes of a stranger.

When we turned into Short Reach we passed a longliner outbound for the fishing grounds. I knew all three of the men who stood in her wheelhouse, yet none of them waved to me, and I did not wave to them.

As we approached the cove, a jet of white mist shot upward, hung for a moment and dissolved, as the Guardian's long back slipped beneath the waters. His presence was final proof that the imprisoned whale had not escaped, in the flesh at least. He was down a long, long time and when he rose he lingered for a while upon the surface, motionless, it seemed. I am sure he was listening . . . listening for a voice he would not hear again.

The R.C.M.P. launch was in the Pond when we entered and together we searched. Although the waters were so calm and crystalline that we could scan the bottom to four fathoms, the deeper reaches were too dark to penetrate. We could not look into the mystery where she lay.

I do not know, can only wonder, why she did not die with her head upon the shore. I can only guess that, in the darkness of her dying, something in her weary mind willed her to seek the deeps, the lightless ancient womb of ocean.

When we had given up our fruitless search, the launch and the dory came together and lay idly in the middle of the silent Pond.

"Do you think there's any chance she might have got away?" Constable Murdoch wondered.

I knew his question was asked out of innocence and out of hope, and so I repressed a sharp reply and only shook my head. It was Danny who answered.

"Don't be so daft. She's laying in nine fathom right under your feet. In three, four, days she'll blast and come back on top. And won't that be something for the sports to crow about! Aye, eighty tons of rotten blubber to remind them what heroes they is."

Then Danny turned to me. His lean, sardonic face was, as usual, almost expressionless; but when he spoke, the scorn had faded from his voice . . . almost.

"Don't rightly know who was the foolishest . . . them fellows and their gunning, or you, Farley, me son. The way I sees it you done that whale no good. You done Burgeo no good. And I don't say as you done yourself much good." He looked directly at me, and I had nothing to say. He shook his head. "Ah, well, t'hell with it."

Chapter 20

JOE LEADS MOURNING
FOR ILL-STARRED
FINBACK MOBY JOE

St. John, Nfld. c.p.

Moby Joe, apparently dead, was eulogized in the legislature today.

The tale of the trapped Finback whale in a salt-water pond near the South Coast fishing community of Burgeo ended earlier in the day when it disappeared, believed to have died and sunk.

Premier Joseph Smallwood, benefactor of the 80-ton creature, rose at the opening of the sitting to announce the news.

He said: "I'm sure that all of Newfoundland, all Canada and even all North America, will hear this news with regret."

. . . The battle to save the whale ended today in defeat when it succumbed to what experts believe was a massive infection resulting from gunshot wounds ten days ago . . . it is presumed to have died during the night and sank in nine fathoms of water.

Wednesday, February 8th, began in normal Burgeo fashion. At dawn it was blowing a gale of wind from the sou'east. Clouds streamed so low that the black islands were almost lost to view. Stinging slants of driving snow beat like whip-lashes amongst the scattered, silent houses of Messers Cove. Then, all at once, the storm abated. The wind still blew, but gently, over a cleansed world shining under a pallid sun. In her journal Claire wrote:

"How sad we are on this bright and lovely day. Burgeo looks so beautiful, and I don't care anymore. Our whale is gone. We sat together and listened to the radio news accounts and I could not stop thinking about the savage mentality of the men who stood around the Pond and emptied their rifles into that huge and harmless animal. Surely *they* are the beasts, not the whale.

"Now it is over. Farley and I are alone with ourselves, having to face the depressing reality of what life will be like if we stay on here. . . ."

It was a lovely day . . . but one of the loneliest I can remember. There was no meeting between us and the people of Burgeo. Nobody called. Not even our closest neighbours crossed the unmarked stretch of new snow around our house. No children came, as they had always come before, to sit quietly in our kitchen after school. I felt as if we were marooned on a cold rock in a lifeless world. Our only human contacts now seemed to be with the thin and disembodied voices of strangers far away, heard over the metallic whimpering of the telephone. And soon, now that the whale was gone, those voices too would fade into silence.

The oppressive sense of limbo grew too much to bear, and in the late afternoon, as the light was fading, I took Albert, our water-dog, for a walk. Deliberately I headed east along the rough path leading toward the main part of Burgeo. I met a few people but, although they returned my greeting with civility, I was not comforted. I was sure they were concealing their real feelings. I became certain of it when I passed a group of young men and women, employees of the fish plant, making their way homeward at the end of the day's work.

They parted to let me by, but had nothing to say until several yards separated us again. Then I heard girls' voices chanting raggedly, and not very loudly:

> *Moby Joe is dead and gone ...*
> *Farley Mowat, he won't stay long ...*

We turned back then, Albert and I, and made our way out to Messers Head from whose lonely summit I had watched the Fin Whales fishing only a few weeks earlier.

It was too dark to see very well. I could barely make out the old stone beacon on Eclipse Island where, almost three centuries earlier, Captain James Cook had taken observations on a transit of Venus.

I sat for a long time on the crest of the Head, locked in the confines of my mind, savouring the bitter taste of my defeat. Then slowly I became conscious of the eternal sounding of the seas, and my thoughts drifted away from myself and the world of men, turning outward to the void of ocean, and the world of whales.

For the first time since the trapped whale vanished, I became fully aware of a rending sense of loss. It was dark, and there was none to know that I was weeping ... weeping not just for the whale that died, but because the fragile link between her race and mine was severed.

I wept, because I knew that this fleeting opportunity to bridge, no matter how tenuously, the ever-widening chasm that is isolating mankind from the totality of life, had perished in a welter of human stupidity and ignorance—some part of which was mine.

I wept, not for the loneliness which would now be Claire's and mine as aliens among people we had grown to love, but for the inexpressibly greater loneliness which Man, having made himself the ultimate stranger on his own planet, has doomed himself to carry into the silence of his final hour.

It was well past supper time when Albert and I came home. Saffron light streamed over the snow from the bay window of the little house. I pushed open the kitchen door . . . and the house was full of people! Uncle Job was there with a drink in his hand and a grin on his weather-scarred old face. There was Sim, and Onie, and several other fishermen. Claire, looking flustered but happier than she had seemed for days, was bringing chairs in from the dining room to accommodate Dorothy Spencer and a number of other youngsters.

Uncle Job raised his glass in salute.

"Foine weather, Skipper. But I believes she be a weather breeder. Aye. Don't say but 'twon't be some dirty day in the marning."

The visitors did not stay long and, as was their custom, they did not say much. But it was what they did *not* say that counted. Nobody spoke one word about the whale.

Sim was the last to leave. He lingered in the door for a moment then, with patent embarrassment, said what he had come to say.

"Don't you fret . . . you and your woman, now . . . you got good friends in this place . . . the foinest kind. . . ."

He could say no more, but he had said what was in his heart and we had understood. Sim Spencer will never know the gratitude we bear him still. The solace of his words was so effective that when, on Thursday morning, some of the fishermen who had worked the capelin seine arrived in a glum, demanding mood, I was able to deal with them without rancour.

They told me they wanted payment, in full and at once, for what they claimed was owing them for their work. It came to more than $500.00. I explained that they would have to wait until the money promised by Premier Smallwood came (not knowing then that it would never come).

"Someone's lying!" one of them replied truculently. "We'se sent off a telegram to Joey, telling as how you and they Sou'westers won't pay we the money he sent."

"Aye," added another. "And we knows you fellows be going to sell that whale; and she worth a fortune, sure!"

This was too much.

"What the devil are you talking about! The whale is dead at the bottom of Aldridges Pond. How do we sell a whale we haven't got, supposing anyone was fool enough to want to buy what's left of her?"

They exchanged uncertain glances and then the spokesman said:

"Well, bye, she's back. She's floating in the Pond. An' we hears they Sou'westers has already sold she to a whaling company from up along Nova Scotia way for ten t'ousand dollars!"

He paused, then added with sudden belligerence:

"Dat money belongs to Burgeo folk. Not to they fellows. No . . . nor to no strangers, neither!"

There was little point in trying to argue against such wild rumours, and I did not want to argue. The whale had risen, and so there were some final things for me to do. I gave each man my personal cheque for the sum he claimed was owed to him. When they had left I called the manager of the fish plant. Had he heard that the whale had risen? He had not. And he was appalled.

"The devil you say! If I'd heard I'd have had her carcass towed out of there and turned adrift at sea before anyone knew a thing about it. Would be the best thing for Burgeo. I'll send a boat and crew to do the job right now!"

"You'd better think about it a bit first," I told him. "That carcass weighs about eighty tons. If you turn it adrift it'll become a menace to navigation. In a collision it could sink a good-sized ship, and you'd be responsible."

"Then what'll we do? You're the keeper of the bloody thing!"

"That's where you're wrong. I *was* the keeper of a

living whale. The corpse is Burgeo's. It belongs to Burgeo, and especially to those who had a hand in killing it. Let *them* look after it!"

Radio bulletins were soon announcing the news of the whale's resurrection and shortly thereafter I had a phone call from the woman doctor, who was also Burgeo's public health officer. According to her, the corpse was extremely dangerous. The infection, if transmitted to a human being, might well prove fatal, she explained. It was imperative that I issue a public warning over the radio telling everyone to stay away from the carcass.

I should have refused, for it was her responsibility, but I did as I was asked; and shortly thereafter all hell broke loose. The plant manager, almost inarticulate with fury, called to demand that I repudiate the warning. He reminded me that the plant drew some of the water for its operations directly from Short Reach, less than a mile from where the dead whale floated. He was obviously frightened that the federal fish inspectors would get wind of the matter and close the plant.

He got no sympathy from me.

"You might have given that some thought a couple of weeks ago, when someone down there was issuing army ammunition to the sports so they could have their bit of fun."

I was being vindictive, but I make no apologies. In my journal notes for that day I wrote:

"The Ancient Mariner had nothing on Burgeo. He only had an albatross slung round his neck. Burgeo has eighty tons of poisonous meat and guts and blubber. Maybe the people here will get the message."

They did, indeed, get a message, but it was not the one I had hoped for. In the face of a renewed outburst from the mainland press and radio, highlighted by a scurrilous article by Bob Brooks in the *Star Weekly*, the majority of the Burgeo people, even those who had been blameless in the tragedy or who had shown sympathy for the whale, now closed ranks. The media attack was a shotgun blast fired at random, and it was only to be expected that many of the residents, hurt, angry, and unable to reply or to defend themselves, should have reacted as they did. Almost overnight the lines became clear-cut. The men who had killed the whale began to be regarded as innocents, if not as victims, and it was widely agreed that they had been justified in doing what they did.

In such circumstances it was inevitable that the Sou'westers Club, composed as it largely was of the town's business and professional men, should change sides again.

Late Thursday evening I had a call from one of them. He was polite, but the gist of his message was that the club was publicly dissociating itself from me and that, from this point forward, it would be concentrating its efforts on restoring the good name of Burgeo.

I wished him luck, and I sincerely meant it.

Not until late Friday morning was the weather good enough to allow Claire and me to pay our last visit to Aldridges Pond. Once again it was a frigid day, with frost smoke rising from the tickles, and cat ice crackling under the bow of Onie's dory. Although it was a fine day for whale spotting, we saw none. There were no distant puffs of vapour hanging like exclamation marks above the dark-skinned sea. The calm waters of Short Reach were unbroken by the swirl of the Guardian's great flukes. The surviving members of the family had vanished, and they were not seen again that winter nor, as far as I have heard, have they or any of their kind again returned to what was once their sanctuary among the Burgeo Islands.

It would have been a lifeless scene save that, far above the still waters of the Pond, three eagles soared on the updrafts over Richards Head. Like silent mourners, they drifted in the void of air above the void of sea. As I stared up at them, I thought how fitting it was that these masters of the aerial world, themselves already doomed, should have chosen this time and place to describe their majestic arabesques . . . For when I dropped my gaze we were in the Pond and there before us floated all that was left of one of the true masters of the sea, a portent of the almost certain extermination of the Great Whale nation.

Even after the lapse of years, I grieve to write about her as I saw her then. She had been immense in life—now she seemed twice as huge. She was floating on her back, high out of the water, and the pallid mountain of her swollen belly was like a capsized ship. From a being of transcendental majesty and grace she had been changed into an abomination; grotesque, deformed and horrible. She stank! She stank so frightfully that, as we cautiously approached her, we had to fight down our nausea.

The most revolting aspect of the scene before us lay in the final confirmation that she was female and, judging from the condition of her breasts, had been far advanced in pregnancy.

I do not know what Onie felt as we drifted beside the monstrous corpse; but Claire was quietly crying. I think we were all grateful for the distraction when the chugging of an engine announced the arrival of another boat.

It was a work boat, manned by three men from the fish plant. It came busily through the channel and, ignoring us, went directly to the whale. The men had handkerchiefs tied over their mouths and noses, giving them a sinister look. They worked quickly to secure a loop of wire cable around the whale's tail just forward of the mighty flukes. Then the work boat, dwarfed to insignificance by her tow, put her stern down and white water foamed under her counter as she took the strain. Slowly, ponderously, the whale began to move. The bizarre cortege slowly drew abreast of us and turned into the mouth of the south channel.

And the great Fin Whale, who had been unable to pass that barrier alive, floated easily over it in death . . . returning, now that there was no return, to the heart of mystery from whence she came.

To Make Amends

Before that most rapacious of predators, the human animal, set about annihilating them in earnest during the 17th century, the eight species of Great Whales are believed to have numbered as many as four and a half million individuals.

By 1930, three centuries later, they had been reduced to about one and a half million.

Less than half a century after that, in 1972, there were estimated to be no more than three hundred and fifty thousand survivors.

One race, the Atlantic Grey, which was known as the Scrag Whale to the New Englanders who destroyed it, became extinct even before science realized it existed. Four others: the Southern Right or Biscayen, the Northern Right or Bowhead, the Blue, and the Humpback, are now so close to extinction that, despite nominal protection, it is doubtful if they will endure. Three species: the Fin, the Sei and the Sperm, now bear the brunt of our unremitting assault on the whale nation and their numbers are melting away with appalling rapidity. Only one race, the Pacific Grey, which has been protected for more than forty years, seems to be tentatively edging *away* from extinction, rather than plunging toward it.

Because of the catastrophic decline in the numbers of Great Whales, many nations which were once prominent in the "industry" have given it up as being

THE PASSING OF THE GREAT WHALES*

	Estimated Numbers Before Intensive Hunting Began	Survivors in 1930	Survivors in 1972
Grey Whale (Atlantic race)	100,000	Extinct	Extinct
Right Whale (Southern or Biscayen)	200,000	Rare	80 to 250
Right Whale (Arctic or Bowhead)	500,000	Rare	100 to 500
Blue Whale	600,000	100,000	600 to 1,500
Humpback Whale	300,000	Rare	1,500 to 2,000
Grey Whale (Pacific race)	100,000	Rare	8,000 to 10,000
Fin Whale	1,000,000	600,000	50,000 to 60,000
Sei Whale	400,000	300,000	60,000 to 75,000
Sperm Whale	1,200,000	600,000	150,000 to 200,000
	4,400,000	1,600,000	270,280 to 349,250

of no further economic interest. These include the United States, Great Britain, Holland and Germany. In 1972 only Japan and the u.s.s.r. continue pelagic whaling on a major scale, although Norway—by all odds the most successful whaling nation in human history—is snapping at their heels in this last paroxysm of slaughter. Japan and the u.s.s.r. between them take about 88% of the forty to forty-five thousand Great Whales, which, under the quotas set by the International Whaling Commission, may be legally butchered. The remainder of the "legal" kill is mainly taken by shore stations under Canadian, Norwegian, Japanese and South African control.

The official figures issued by the I.W.C. are bad enough—but they do not, by any means, tell the whole story. Most members of the I.W.C. (and several whaling countries are not even members) routinely fail to report the "accidental" killing of undersized whales, cows with calves, and whales of "protected" species, as well as direct quota violations. Since the I.W.C. does

---

*These figures have been compiled from a careful examination of all available sources. The estimates for the period before the beginning of intensive whaling are probably conservative; those for the surviving numbers of Great Whales in 1972 probably err on the side of liberality.

not provide either effective surveillance, or meaningful sanctions against offenders, the quotas which it sets have always been, and continue to be, as much honoured in the breach as in the observance. To make matters worse there is an increasingly large number of catcher vessels (and even some small catcher-factory ships), mostly operated by Japanese and Norwegian nationals under flags of convenience, whaling on the high seas with a total disregard for *any* regulations or conventions. These "pirate whalers" are known to take whales of any size, sex or species—including *all* of the nominally protected species—whenever and wherever they can. Sometimes operating with the connivance of maritime states, particularly in South America and southern Africa, they are believed to kill between two and five thousand Great Whales every year, *none of which* appear in "official" statistics of the I.W.C.

About 50,000 of the larger whales will die in 1972 at the hands of men. Yet, although such continuing slaughter must result in the virtual extermination of most species of whales before the end of the present decade, *no action to significantly reduce the size of the annual kill is being taken by the nations most involved in whaling.* As has been the case since its inception, the International Whaling Commission was, as late as 1971, still setting annual quotas for each species which were so unrealistic as to sometimes be *in excess of the numbers of such whales that the ships and shore station of the member states could handle.* This is not conservation . . . it is legitimatized mass destruction.

The Great Whales are not alone in being hustled toward extinction. We are now seriously threatening the survival even of those relatively little baleen whales, the Minkes and the Brydes (now being actively hunted by catchers from Norway), as well as the White Whales (Beluga) and the Potheads (Pilot Whale). Canada permits tourists to shoot Beluga *for sport,* as part of the entertainment offered to arctic trippers! To make matters worse, man is now making purposeful war on many species of dolphins and porpoises. During 1971, more than two hundred thousand of these little toothed whales were taken by the Japanese for commercial purposes, while an almost equal number are believed to have perished "accidentally" in tuna seines in Pacific waters. As the last Great Whales are converted into pet food and cosmetics, the Japanese will certainly intensify the pressure on the dolphins and porpoises although, apparently, the Russians do not intend to follow suit. Mr. A. Iskov, the Soviet Minister of Fisheries, recently prohibited the killing of these small whales, on the grounds that they are "the maritime brothers of mankind." There is even some reason to hope that the Soviets would be willing to reduce, or even stop, their part in the slaughter of the last of the Great Whales—*if* the Japanese could be persuaded to do likewise. So far the Japanese, whose representative on the I.W.C. is also a director of a major Japanese whaling company, have adamantly refused to cooperate.

It is abundantly clear that if we are to save *any* of the whales, Great or Small, we must reject the I.W.C. as our instrument for preventing the ultimate commission of a crime against life which is of such magnitude that it has no equal in human history. The I.W.C. has never served the whales . . . it has only served the whalers.

If the whales are to survive, mankind must declare *and enforce* a world-wide moratorium on the killing of all and any whales. Such a respite must be of at least ten years' duration if the terribly depleted stocks of Great Whales are to have any real chance to recover. During the moratorium there must be a strictly enforced world-wide embargo on trade in all whale products, otherwise many whaling companies will simply transfer their operations to flags of convenience so that their ships may join the growing fleet of pirate whalers.

As this book goes to press it appears that a modicum of progress is being made toward saving the remaining whales. The United States has unilaterally declared an embargo on the importation of oil, meat, or any other products derived from the eight species of Great Whales. On June 9, 1972, a resolution was passed by fifty-one of the nations attending the United Nations Conference on the Environment, in Stockholm, calling for a ten-year ban on commercial whaling. Although this appears, at first glance, to be a major victory for the whales, it is by no means sure that it will amount to much more than a gesture. It will surely be no more than that if, as has been recommended, the I.W.C. is given the responsibility for arranging such a moratorium. Although a two-thirds vote by the members of the I.W.C. might result in nominal acceptance of a ban on whaling, such a ban would only be binding on the member states on a *voluntary* basis, and any member could simply withdraw from the I.W.C. and continue whaling as it chose—something that has happened in the past. On the other hand, it is by no means certain that the recommendation of the Conference on the Environment will ever come before the General Assembly of the United Nations; but, even if it does, and even if the General Assembly were to vote in favour of the moratorium, it would still not be binding on any nation which chose to reject it. Consequently, nothing short of a full-scale international treaty can provide the protection which is so urgently needed. It is therefore imperative that the various governments which are in any way involved in whaling, or with the use of whale products, be subjected to a sustained and insistent pressure demanding their participation in a legally binding treaty of international scope.

A number of organizations are strenuously supporting the need for such a treaty. One of the leaders among these is Project JONAH, which operates under the auspices of Friends of the Earth. JONAH welcomes inquiries as well as offers of assistance. It maintains offices at the following addresses:

Project JONAH   Project JONAH
Box 476   25 Quai Voltaire, Paris 7e
Bolinas,   France.
California 94924
U.S.A.

Some additional organizations concerned with the preservation of whales, porpoises and dolphins, are:

Fauna Preservation Society (U.S.A.)
International Society for the Protection of
    Animals
International Union for Conservation of Nature
    and Natural Resources
New York Zoological Society (U.S.A.)
The Environmental Defence Fund (U.S.A.)

Individuals who wish to make their voices directly heard on behalf of whales and of the moratorium may write to the International Whaling Commission, care of its President:

Hon. J. L. McHugh,
Marine Science Research Centre,
State University of New York,
Stony Brook, N.Y. 11790,
U.S.A.

Citizens of the United States can also make their feelings known to:

The Secretary of the Interior,
Department of the Interior,
Washington, D.C.
U.S.A.

and Canadian citizens should express their wishes to:

The Minister,
Department of the Environment,
Ottawa, Ontario.

Letters requesting the cooperation of the governments of Norway, Japan and the U.S.S.R. in establishing a moratorium under international treaty should be addressed to the Ambassadors of these nations at their embassies in the country of the letter writer's origin.

If we are to make amends to the whale nations for the despicable savagery with which we have treated their members in the past, we must do so now. In a few more years there will be nothing left that we can do for them ... except to write their epitaph.

June 15, 1972

Commission rejects whaling moratorium

LONDON (Reuter) — A U.S. call for a 10-year halt to all whale hunting was rejected yesterday by the International Whaling Commission.

The U.S. plea for a moratorium was thrown out by a plenary session of the IWC's 24th annual meeting. The vote was 6 to 4 with four abstentions.

The moratorium idea was put forward amid a climate of emotional protest from conservationists that some of the earth's largest creatures were close to extinction.

The meeting was barred to press and public. But informed sources said scientists of the major whaling nations argued that the species now being hunted were numerous enough to survive.

The U.S. moratorium proposal was seconded by Britain and supported by the Argentine and Mexico. Against it were Japan, the Soviet Union and Norway—who conduct the great bulk of the world's deep-sea whaling—plus Panama, South Africa and Iceland. Australia, Canada, Denmark and France abstained.

The Globe and Mail
June 30, 1972

ABOUT THE AUTHOR

Farley Mowat, author of such distinguished books as *People of the Deer*, *Never Cry Wolf*, *A Whale for the Killing*, *The Snow Walker*, and *Sea of Slaughter*, has long been eloquent in his indictment of man's exploitation of human and non-human life on this planet. He was born in Belleville, Ontario, in 1921 and began writing for a living in 1949 after spending two years in the Arctic. He has lived in or visited almost every part of Canada and many other lands. More than ten million copies of Farley Mowat's books have been translated and published in hundreds of editions in over forty countries.